Planet Earth
UNDERGROUND WORLDS

This volume is one of a series that examines
the workings of the planet earth, from the
geological wonders of its continents to the
marvels of its atmosphere and ocean depths.

Cover
A multiple-exposure photograph details the
progress of a caver descending the Incredible Pit,
a 440-foot natural shaft in Ellison's Cave
in Georgia. Dangling precariously from a nylon
rope and drenched by an underground waterfall,
the caver triggered a hand-held flash four
times to record the pit's awesome depth.

Planet Earth

UNDERGROUND WORLDS

By Donald Dale Jackson
and The Editors of Time-Life Books

Time-Life Books, Amsterdam

PLANET EARTH

EDITOR: Thomas A. Lewis
Designer: Donald Komai
Chief Researcher: Pat S. Good

Editorial Staff for *Underground Worlds*
Picture Editor: John Conrad Weiser
Text Editors: William C. Banks, David Thiemann
Writers: Tim Appenzeller, Adrienne George,
John Newton
Researchers: Judith W. Shanks (principal),
Susan S. Blair, Therese A. Daubner, Stephanie Lewis,
Barbara Moir, Donna Roginski
Assistant Designer: Susan K. White
Copy Coordinators: Allan Fallow, Victoria Lee,
Bobbie C. Paradise
Picture Coordinator: Donna Quaresima
Editorial Assistant: Annette T. Wilkerson

Special Contributor: Champ Clark

Correspondents: Elisabeth Kraemer (Bonn); Margot Hapgood, Dorothy Bacon (London); Susan Jonas, Miriam Hsia, Lucy T. Voulgaris (New York); Maria Vincenza Aloisi, Josephine du Brusle (Paris); Ann Natanson (Rome). Valuable assistance was also provided by: Mirka Gondicas (Athens); Pavle Svabic (Belgrade); Helga Kohl (Bonn); Millicent Trowbridge (London); M. T. Hirschkoff (Paris); Mimi Murphy, Anne Wise (Rome); Traudl Lessing (Vienna).

THE AUTHOR

Donald Dale Jackson is a former staff writer for *Life* and the author of other Time-Life books in The World's Wild Places and The Epic of Flight series. Among his other publications are a volume on the judicial system of the United States and a history of the California gold rush.

THE CONSULTANTS

Dr. John Holsinger is Professor of Biological Sciences at Old Dominion University in Norfolk, Virginia, and a Fellow of the National Speleological Society. A recognized specialist on cave invertebrates, he has published numerous articles on cave fauna.

Ernst Kastning, Assistant Professor of Geology at the University of Connecticut, has written extensively on the history of cave exploration and cave science. He is a Fellow of the National Speleological Society and has served as Director of Publications for the Cave Research Foundation.

Arthur N. Palmer is Professor of Geology and Director of the Water Resources Program at the State University of New York. He has spent more than a quarter of a century exploring, mapping and interpreting the geology of caves and karst regions throughout the European and North American continents.

© 1982 Time-Life Books Inc. All rights reserved.
Fifth European English language printing 1988.

ISBN 0 7054 0745 4

TIME-LIFE is a trademark of Time Incorporated U.S.A.

CONTENTS

"My hair stood on end, my teeth chattered, my limbs trembled," declared one of literature's most famous cavers as he entered a shaft and began the epic adventure recounted in Jules Verne's *Journey to the Center of the Earth*. Although Verne's tale is entirely fictional, he could not have captured better the misgivings many cavers feel as they step into dark and mysterious subterranean worlds. Yet apprehension is mixed with heady anticipation: Caves offer their visitors spectacular scenery, scientific marvels and, above all, the rare exhilaration of discovery.

Caves are the result of minute incremental etchings of a variety of natural processes that, over a period of many millennia, have hollowed cavities extending for scores of miles into the earth's interior. Passages wider than an interstate highway may feed into fissures that are breathlessly thin. Vertical shafts, often obscured by pitch-black shadows, plummet hundreds of feet. Yet here also are found some of nature's most exquisite artifacts. Beneath massive layers of limestone uplifted by some titanic convulsion, explorers may find a tiny shimmering crystal more fragile than spun glass.

By scrutinizing rock formations in caves, scientists have found clues to geologic events that occurred a billion years ago. And by studying the remarkable creatures that thrive in a cold, wet world of impenetrable darkness, these explorer-scientists have illuminated the wondrous adaptability of living things.

But as much as caves have given to science, they remain chiefly the domain of amateur explorers. Every year, thousands of cavers probe the underground frontier, seeking passages and chambers such as the ones shown here and on the following pages. Guided by small lights and a keen sense of adventure, they thread their way through perilous mazes with a persistence that confirms Jules Verne's observation: "There is nothing more powerful than this attraction toward an abyss."

A team of cavers exploring the nine-mile-long labyrinth of St.-Marcel d'Ardèche in southeastern France pause to survey the grandeur of a central gallery. The cave, carved out by the Ardèche River system tens of thousands of years ago, still offers tantalizing opportunities for discovering unknown passages.

7

With two waterproof flashlights strapped
to his helmet, a diver emerges from the water
that almost fills a gallery in the Cave of
Caumont in northern France. Diving equipment
is so cumbersome that cavers take it along

Struggling through the aptly named Agony
Crawlway in Ellison's Cave in Georgia,
an explorer hauls thick coils of strong rope to
be used for a pit descent. Not far beyond
The Agony lies The Fantastic Pit—at 510 feet,
the longest vertical drop in U.S. caves.

Suspended by one rope and steadied by a second, a caver lowers himself into a deep pit in San Agustín Cave in Mexico. Explorers routinely negotiate dizzying vertical shafts, using mountain-climbing techniques.

In Phanonga, Thailand, a series of limestone
terraces totaling 20 to 30 feet in height
supports a delicate shroud of flowstone. The
formation is the result of mineral-laden water
spilling over the terrace edges, depositing
microscopic particles with each tiny surge.

A 20-foot column surrounded by rapier-like
stalactites connects the floor and ceiling of the
Dome Room in New Mexico's Carlsbad
Cavern. Such columns form when stalactites
above meet stalagmites on the floor—a
process that may take 100,000 years.

Deep within Mammoth Cave in Kentucky, a caver pauses to examine slabs of rock that have fallen from the cave ceiling. Because it signals the end of the cave's life cycle, this collapse of rock strata is called cave breakdown.

EXPLORERS OF A STYGIAN REALM

Andy Eavis, a 30-year-old professional engineer employed by a Yorkshire colliery, discovered the first clue to the existence of a major cave in Borneo in late 1977. At the time, he was sitting comfortably at home in Selby, England, perusing aerial photographs of a rain forest on the other side of the world. Eavis was also the chairman of the International Speleological Union's exploration section, and the Royal Geographical Society had invited him and five other cave experts to join an expedition into the newly designated Gunong Mulu National Park in the Malaysian state of Sarawak on northern Borneo. While more than 100 scientists—including botanists, zoologists, ecologists, geologists, anthropologists—were to study the pristine rain forests aboveground and prepare a management plan for the park, Eavis and his comrades would probe its subterranean honeycomb of caves.

The photographs of Mulu's unmapped jungle, which was populated only by nomadic Penan tribesmen, were dominated by a jagged, three-mile-wide limestone ridge that extended some 20 miles in a northeast-southwest direction in north Sarawak. The ridge, whose peaks reached as high as 6,000 feet, was a perfect example of what geologists call karst, the terrain formed by deeply eroded limestone. Most of the erosion is done by water; rather than shedding rain water, karst absorbs it like a sponge. For two million years heavy tropical rains had been sculpting the limestone into a strange, craggy landscape where rivers suddenly vanished into the ground, yawning sinkholes opened in the earth and rock bridges spanned whole canyons.

It was obvious to any experienced caver that this terrain concealed a labyrinth belowground. After surveying a few easily accessible Sarawak caves for the Malaysian government in 1961, geologist Gerald E. Wilford had reported that "large spectacular caves are most likely to be discovered in the uninhabited and relatively unexplored Melinau area between the Tutoh and Limbang Rivers"—a region within the 210-square-mile park.

Eavis and his colleagues called themselves simply cavers (cave cognoscenti detest the term "spelunker," although scientifically inclined cavers do style themselves speleologists), but they were scarcely amateurs; the Royal Geographical Society had vetted both the caving experience and scientific credentials of these six before inviting them to join the expedition. Team members boasted professional expertise in geology, geomorphology, cave biology and hydrology. Among them, they had mapped and studied every major cavern in Europe, descending shafts hundreds of feet deep and spending days underground. Their skill in rock-climbing techniques—belaying, prusiking, rappelling and the like—surpassed that of most mountaineers.

By the eerie glow of torches, members of an 18th Century expedition inspect a cave near Cornial, in what is now Yugoslavia. The group, led by Austrian explorer Joseph Nagel, had been ordered to map the limestone caves of the area by Holy Roman Emperor Francis I.

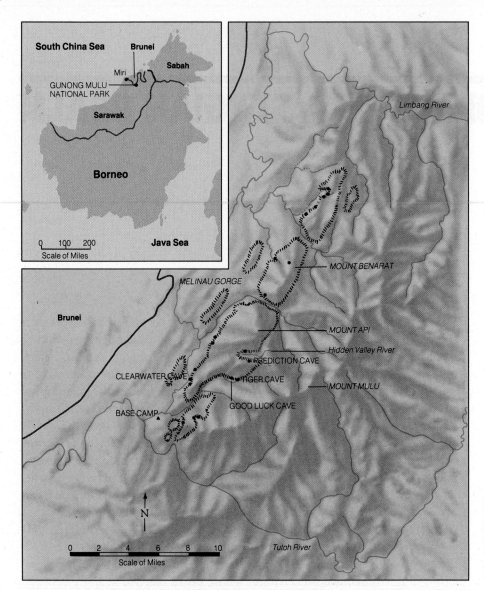

Gunong Mulu National Park, site of extensive British caving expeditions in 1978 and 1980, is located in Sarawak in northwestern Borneo *(inset)*. The explorers concentrated their efforts on a 20-mile-long, three-mile-wide limestone ridge system *(hatched outline)* that contains more than a dozen caves *(dots)*.

As he studied stereoscopic photographs through a special viewer, Eavis spotted any number of cave entrances in the Melinau Limestone, the geological formation that made up part of the ridge that interested him. He was particularly intrigued by a huge limestone overhang on one of the ridge's peaks above a river that seemed to abruptly sink into the ground. Because the overhang was above a deep valley, Eavis knew that a prehistoric river many times the size of the present stream had once flowed through the valley and under the cliff, dissolving the limestone beneath the overhang—and perhaps leaving an enormous cave along its subterranean course.

The hypothesis remained in the back of his mind during the ensuing months, as the cavers attended to the details involved in mounting any wilderness expedition: immunizations, solicitations of food and equipment, travel arrangements, fund raising. They arrived at their Mulu National Park base, about 60 miles southeast of Miri on the southern end of the limestone ridge, in March 1978. Local porters and tribesmen had assured expedition members that there were no caves in the region, but the cavers decided to pursue Eavis' hunch even though it would be an arduous gamble.

Eavis and two colleagues were guided to the site of the putative cave by the leader of the scientific expedition, Robin Hanbury-Tenison, who had already scouted the area. With four porters, they slogged northeastward for two days through the knee-deep muck of a steamy flood plain, quickly becoming accustomed to the jungle's hazards: a debilitating combination of

85° F. heat and 100 per cent humidity; constant assaults by buzzing clouds of mosquitoes and sand flies; encounters with leeches, centipedes, scorpions and king cobras; and not least of all the ravages of *borak*, a potent home-made rice wine forced on the cavers by friendly tribesmen.

Toward the end of their 10-mile trek they forded raging rivers in the foothills and finally climbed the knife-edged ridge, 3,000 feet high at this point, on the southern end of Gunong Api, or Mount Api. They then descended into the valley in which the overhang was located, dubbed it Hidden Valley, and set up camp. Because their group was too small to mount a rescue operation should disaster strike, they set up short-wave radio communications with the base camp. Next day, Eavis and the other two cavers scrambled several hundred feet up the precipitous valley wall through rocky debris and thick vegetation to the massive overhang. Under it they found a steep slope, slick with mud and water, plunging 600 feet into the mountain. There the entrance opened into a soaring overhead shaft 200 feet in diameter and perhaps 1,000 feet high. The team named their discovery Lubang Ramalan—Prediction Cave—in honor of Eavis' shrewd prophecy.

Exploring the cave beyond the shaft was a sweaty, risky ordeal. The jagged limestone could slice like a razor through cloth and flesh, and the standard safety procedure was to wear protective coveralls. Struggling along in the tropical heat was, according to one caver, "like having an airless, hot shower." Torrential rains frequently generated subterranean flash floods that could trap or drown cavers. Even dried bat guano on the cave floor could be dangerous: It harbored fungus spores that cause histoplasmosis, a potentially deadly fever.

To all these hazards Prediction Cave added special obstacles. The passage narrowed to a crawlway that was 300 feet wide but only two feet high, its bottom littered with pebbles and wet clay. It was, one caver wrote later, "one of the most ridiculous passages under the earth." For a seemingly interminable time, the cavers slithered on their bellies through this black crack, pulling their gear behind them in canvas bags, guided only by their flickering carbide helmet lamps. After struggling thus for 400 feet they reached a crumbling chamber nearly filled by an 80-foot mound of boulders. The cave seemed to continue into the mountain on the other side of the mound—a trench extended in that direction—but the cavers were at a dead end. Another pile of fallen boulders, a choke, filled the passageway.

The explorers did not yet despair, but instead resorted to geological detective work to reconstruct the cave's history. A huge river apparently had been flowing through the cavern when the roof collapsed and restricted its flow with the boulder choke. The slower-moving water deposited sediment more rapidly, eventually filling the giant tunnel to its roof. During eons of time the river carved its way deeper into Hidden Valley, the water receded from the cave and the sediment gradually settled, creating a cave floor that bore an imprint of the contours of the roof. By this reasoning it seemed likely that a passage—perhaps an enormous one—lay beyond the choke.

The ancient river must have flowed out of the mountain somewhere, and the cavers scoured its flanks for days but did not find the missing link to Prediction Cave. Instead, a few miles to the southwest, they discovered two other caverns, Tiger Cave and Clearwater Cave, both with impressive streams flowing from their mouths. If these streams were resurgences of the river in Hidden Valley, the existence of a large cave network in the ridge—although not necessarily the one they had entered at Prediction Cave—would be confirmed. To find out, the cavers watched the water emerging from Clearwater Cave while porters dumped 44 pounds of a fluorescent green dye into the river near Prediction Cave. The observers at Clearwater

Mulu: The Genesis of a Cave Region

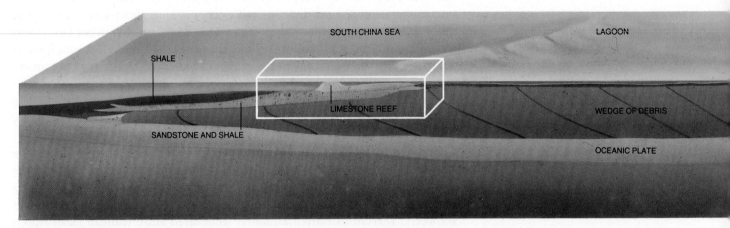

SHALE

SOUTH CHINA SEA

LAGOON

LIMESTONE REEF

SANDSTONE AND SHALE

WEDGE OF DEBRIS

OCEANIC PLATE

The magnificent caves of the Mulu region in northwest Borneo exist because of a remarkable sequence of gentle sedimentation and titanic collisions that began some 60 million years ago. At that time, one of the earth's great slabs of crustal material, or tectonic plates, began to dive under another plate carrying the land mass now known as Borneo.

As the plate ground downward, thick layers of sediment were scraped off and piled up at the edge of the land mass. This wedge of debris accumulated slowly beneath the sea and year after year was blanketed by the remains of billions of marine animals. This ages-long submarine blizzard of calcite sea shells created thick drifts of new sediment, some of which were eventually compressed into a limestone barrier reef on top of the wedge of debris (above). Concurrently, mud and clay on the surrounding deep-ocean floor consolidated into shale.

Millions of years later, further move-ments of the plates caused this entire area to sink, and the limestone was buried by layers of shale and other sediments (right, top and middle). About five million years ago, however, the tectonic dynamics changed; the diving plate rose and began to grind more directly against the overriding plate. The awesome energies of this collision buckled and lifted the rock layers at the edges of the plates until a new shale-capped mountain was thrust above the sea.

After two million years, heavy tropical rains had worn off the shale, the highly soluble limestone was uncovered and erosion accelerated both on the surface and deep within the rock. During the past million years, erosion whittled the mountain into three separate peaks (right, bottom) that dominate the Mulu landscape today: Mount Mulu, made of sandstone and shale, and the twin limestone pinnacles of Api and Benarat, which house the spectacular Mulu caves.

An artist's conception of the conditions 40 million years ago shows in cross section an area of the northwest coast of present-day Borneo. The oceanic plate carrying the South China Sea floor is being subducted beneath Borneo, and layers of sedimentary rock, including a limestone reef, are forming offshore.

BORNEO

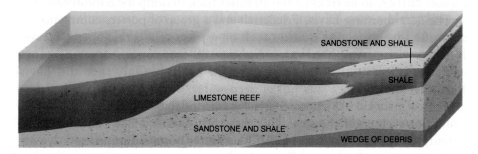

SANDSTONE AND SHALE

SHALE

LIMESTONE REEF

SANDSTONE AND SHALE

WEDGE OF DEBRIS

A close-up of the area where the Mulu caves would emerge portrays the geological situation about 20 million years ago. The limestone reef has subsided beneath the sea and is buried under shale. Sediments of coarser-grained sandstone and shale have begun to wash into the area from inland Borneo to the southeast.

The stresses exerted by the moving plates, accompanied by earthquakes, cause irregular folding of the layers of shale and sandstone. By 10 million years ago, the younger sediments from Borneo blanketed the entire reef complex.

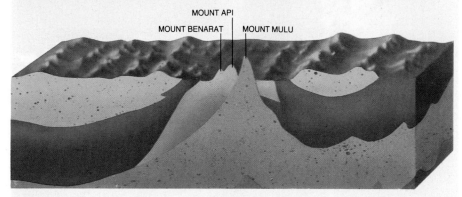

MOUNT API

MOUNT BENARAT   MOUNT MULU

Continued uplifting raises the reef complex high above sea level. It took about two million years for the shale cap of the mountain to erode; in the ensuing one million years, the limestone eroded so fast that the pinnacles of Api and Benarat, where the caves are located, are now about 2,500 feet lower than the 8,000-foot sandstone-and-shale peak of Mount Mulu.

the pair strung safety guidelines across tricky sections until, perhaps 400 feet into the traverse, they had paid out their entire supply of rope. They faced a smooth, blank wall that demanded safety lines; reluctantly, they returned to the cave entrance.

The next day Eavis and Friederich carried large coils of extra rope to the canyon, clipped themselves to the safety lines with metal carabiners and rapidly traversed the explored section, then laboriously clambered spider-like across the sheer face, hammering metal bolts into the rock to secure the safety line. Just beyond, the canyon widened, and they returned to the river, which broadened and soon disappeared: It was emerging from the cave floor at the base of a steep uphill slope. As the pair began climbing the slope, another caver caught up with them, shouting that a deluge outside was beginning to flood the canal. The three men hastily backtracked yet again to the surface. Although the torrent quickly subsided, days of delay had exhausted the team's food supply, and they were forced to return to the base camp.

On the following day Eavis broadcast an enthusiastic description of the new cave over the short-wave radio network that connected the expedition's far-flung outposts. He estimated that the cave was 200 feet wide—not the largest passage in Mulu, but quite impressive. Within 24 hours the base was swarming with speleologists from other camps, all clamoring to explore Good Luck Cave. The dinghies could hold only three cavers with their bulky gear, so the assembled company drew straws for the right to accompany Eavis. The honor was won by Dave Checkley, a veteran cave biologist, and Tony White, regarded as one of the world's premier cave explorers.

The three cavers quickly retraced the route through the Plunge Pool and the 650-foot traverse, then began climbing the steep, steadily widening slope, which was covered with sandy sediment, scree and boulders. As they ascended, they hastily surveyed the cave, since the expedition rarely had the time or manpower for a separate survey after exploration was complete. From each starting point one caver clambered uphill to the end of a 30-meter steel tape while another measured direction with a compass and the elevation angle with a clinometer; the third took notes and paced off the approximate dimensions of side passages and other landmarks.

At the top of the 1,300-foot slope, about three quarters of a mile from the cave entrance, the voices of the cavers began to echo—and the echoes continued for about 10 seconds before the absolute stillness returned. Unwittingly, they had stepped from the passageway into a large chamber, and had lost contact with its walls. They decided to follow a straight compass course until they reencountered a wall. When they found one, 320 feet due south, it was made of shale. That meant they were now under the Melinau Limestone and had found geological proof that the cave was at the bottom of the limestone layer, on the impermeable bed of the prehistoric river.

Surveying the undulating wall proved exceedingly difficult. The steep uphill climb was littered with huge limestone blocks bigger than houses, some so precariously balanced that they shifted when touched. "We often started surveying round a boulder, thinking it was the passage wall, only to find that the real wall was one hundred feet behind it," recalled Eavis. "Then we'd have to go back and resurvey."

After 12 hours and 77 survey legs of 30 meters each, the trio reached a branch passage barely 100 yards long. At its end they found a boulder choke and felt a strong draft blowing out from it—a sure sign of a connection to the surface. The trench on the north side of the passage and the size and shape of the boulders matched the features Eavis had seen two years before in Prediction Cave. Here, surely, was the missing link to Hidden Valley.

Barely discernible against the glare of the powerful lights they are holding, members of the British exploration team manage to illuminate only a portion of the huge Sarawak Chamber. One team member later described his awe at "hearing the incredible echo and feeling the immensity in every direction."

Expedition members clamber around a boulder to measure a segment of Sarawak Chamber. Surveying the 2,300-by-1,300-foot cave floor, with its profusion of house-sized boulders, took 16 hours of uninterrupted work.

edge of the glow cast by torch or candle, the unfamiliar and alarming sounds caused by the slap and spill of distant water echoing down rocky corridors, the deep pits that lay in wait for the unwary—this was a world of intimidating obstacles that most people were content to avoid.

The few who refused to be daunted usually had purposes other than adventure. Entrepreneurs probed several German caves during the 16th and 17th Centuries in search of unicorn horns, which were coveted for their purported medicinal value. Fortune hunters who thought they had succeeded in their quest unwittingly hauled to the surface the fossil remains of hyenas, bears and other animals that had died in their cave lairs. The effect on the patients to whom the resulting potions were given is not recorded.

There were more thoughtful explorations of caves during the 17th Century: for instance, those of John Beaumont, a Somerset surgeon and an amateur student of mining and geology. When in 1674 lead miners excavating a shaft in the Mendip Hills accidentally pierced a dome-shaped chamber, as sometimes happened in limestone country, Beaumont hastened to the site and hired six miners to accompany him into the cave. Carrying candles, the company descended the 60-foot shaft to the first chamber, which Beaumont proceeded to measure: It was 240 feet long, 8 feet wide and 30 feet high. "The floor of it is full of loose rocks," wrote Beaumont in his subsequent report to the Royal Society, England's most prestigious scientific organization. "Its roof is firmly vaulted with limestone rocks, having flowers of all colours hanging from them which present a most beautiful object to the eye"—apparently the same glittering stalactites and curtains of flowstone that awe visitors today.

The intrepid surgeon then led a 300-foot crawl through another narrow, rock-strewn passage, halting at the brink of a cavern so vast, Beaumont reported, that "by the light of our candles we could not fully discern the roof, floor, nor sides of it." Although the miners were familiar with the underground, they adamantly refused to descend this chasm, even for double pay. So Beaumont went down himself: "I fastened a cord about me, and ordered them to let me down gently. But being down about two fathom I found the rocks to bear away, so that I could touch nothing to guide myself by, and the rope began to turn round very fast, whereupon I ordered the miners to let me down as quick as they could." He landed dizzy but safe 70 feet below, on the floor of a cavern 115 feet in diameter and nearly 120 feet high, where he found large veins of lead ore.

Beaumont also visited nearby caves and returned several times to Lamb Leer, as the first cave was known, to supervise the digging of a horizontal shaft that revealed another large chamber. But his published accounts failed to stimulate much curiosity about caves. After a brief flurry of lead mining, Lamb Leer was abandoned; its entrance shaft eventually collapsed, leaving no trace, and the cave's very existence was virtually forgotten until the chamber was discovered anew in 1880.

Although investigations similar to Beaumont's took place elsewhere in Britain and on the Continent, exploring caves remained an eccentric, hit-or-miss pursuit. The early explorers rarely published reports on their work, so subsequent investigators could not build on old discoveries and techniques. The sole exception to this spasmodic pattern emerged in a region of Slovenia called by its Austrian overlords the Karst. (The name, referring to a 100-mile-wide limestone belt running along the coast of the Adriatic Sea, became the generic term for cave country in the 19th Century.)

One of the earliest Slovenian explorers was Baron Johann Valvasor, a well-traveled nobleman and amateur scientist who lived about 50 miles northeast of Trieste. During the 1670s and 1680s, Valvasor visited 70 caves

Between 10,000 and 40,000 years ago when the Great Ice Age was finally relaxing its grip on the earth, a new breed of human emerged who enjoyed all of the mental and physical attributes of modern man, including an unprecedented artistic ability. Known as Cro-Magnons after the site in France where their remains were first discovered in 1868, these beings left as their principal legacy a variety of vibrant paintings that have been preserved through the millennia on cave walls and ceilings.

Both moisture and the expansion and contraction caused by changes in the temperature cause pigment to deteriorate; thus the best-preserved paintings have been found in the driest and most remote regions of caves, where the temperature is constant. Why prehistoric hunters, who did not live in the depths of caves, ventured there to create their art is a mystery. Some archeologists believe that the paintings were a form of hunting magic; others think they were ancestral totems or fertility symbols. The answer may never be known.

Sadly, the very discovery of these prehistoric treasures has hastened their destruction. Microorganisms introduced from the outside world encourage the development of algae and bacteria on the cave murals. Already, such pollution has dulled many ancient paintings and, in some cases, inflicted a greater amount of damage in just 10 years than the cave's natural environment had caused in thousands of years.

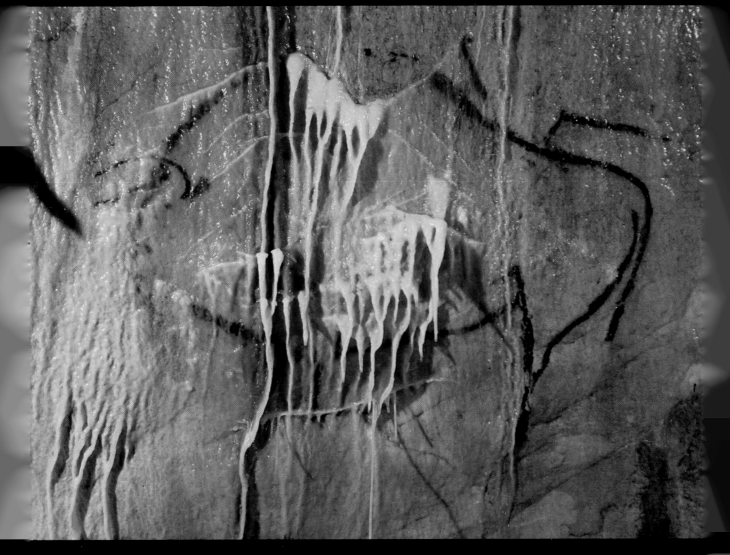

The profile of a bison, outlined by a cave artist 14,000 years ago in Niaux Cave in Ariège, France, still evokes the bulk and power of the ancient animal. Centuries of flowstone accumulation have partially obscured the sketch.

Adorned with silhouettes of the artist's
hands, a prehistoric frieze discovered in Pech-
Merle Cave in southern France features a
pair of dappled horses standing back to back.

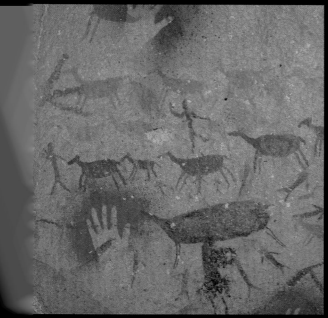

Sticklike human figures appear to be herding
guanacos, wild relatives of llamas, in this
prehistoric Indian painting found in a cave in
Santa Cruz, Argentina. The handprints—a
motif seen frequently in cave art—were made
by placing the hand against the wall and
using a tube to blow pigment over it.

An enigmatic pictograph, including a series
of concentric circles, adorns a cave near
Bustamante, Mexico. Early Indian tribes left
such markings throughout North America.

A large wild cow leaps toward a lattice-like design, possibly representing a trap, in this mural in Lascaux Cave in southern France. Beneath the cow, three small horses gallop away.

This seven-foot-long image of a female red deer (a small bison can be distinguished beneath its throat) occupies a corner of the ceiling in Altamira Cave in northern Spain.

in the Karst. He wrote meticulous reports on his discoveries, illustrated with sketches and maps, and eventually published them in a four-volume, 2,800-page collection. The cave dimensions in this document are unreliable. Valvasor estimated distances without measuring, and he tended to exaggerate—in one case, maintaining that he had explored nearly six miles of a cave that actually was less than half a mile long. But his work was remarkably comprehensive: He was the first to undertake a systematic series of cave explorations and to investigate the underground flow of water within and between caves. For his research he was elected a Fellow of the Royal Society in London.

Nearly a century would pass before Valvasor's example was emulated by another investigator. In 1747 court officials in Vienna ordered a 30-year-old mathematician named Joseph Nagel to explore and map the major caves of the Austro-Hungarian Empire. The reasons for the assignment remain obscure; conceivably Nagel suggested it himself. In the course of a two-year study, Nagel, accompanied by an Italian artist, mapped and sketched several Slovenian caves. His reports were unpublished, and reposed in obscurity in the royal archives. But while they did little to advance cave science, they apparently won the Emperor's favor, launching Nagel on a brilliant career. He soon was appointed court mathematician, then keeper of the royal scientific collections and head of physical science at the University of Vienna.

Both Valvasor and Nagel devoted much of their attention to the most famous Slovenian cavern, Adelsberg Cave (now called Postojna Jama by the Yugoslav government, which annexed the Slovenian Karst after World War II). Adelsberg's conspicuous entrance, a gaping tunnel with the sparkling Pivka River gushing from its mouth, attracted travelers as early as the 13th Century; one alleged visitor was the Italian poet Dante, who may have been seeking material for *The Inferno*. Vacationing nobles and landowners toured the cave occasionally, and their need for guides, candles and torches spawned a cottage industry for local peasants.

The cave's early fame was attributable more to its convenient location near the main thoroughfare between Vienna and the Adriatic Sea than to its speleological wonders. Its chamber contained a few soot-blackened stalagmites, and the passage seemed to dead-end after some 340 yards, blocked by rock walls and the impassable river. But in April 1818, workmen who were preparing for a visit by the Habsburg Emperor Francis I of Austria discovered a larger, fantastically ornamented chamber with an entrance high on the original cave's wall *(pages 42-49)*.

There is no evidence that the 50-year-old Emperor made the climb to view this new marvel, but a district official named Josip Jeršinovič soon rendered it accessible to all and sundry—for a price. Jeršinovič was among the first to realize that fashionable people would pay to tour a spectacular cave, despite their fears, provided that the path was smooth and well lighted. After installing himself as the first manager and treasurer of Adelsberg Cave, he hired laborers to bridge the river, construct a stairway to give easy access to the newly discovered continuation of the cave and level a path through its two and a half miles of passages. To protect Adelsberg from the depredations of souvenir hounds—and its promoters from loss of profits— he created a commission that regulated every aspect of the cave's operation. A locked gate was installed at the entrance and torches were banned in favor of clean-burning candles and oil lamps. Visitors had to sign a logbook, buy tickets to enter, and hire specially trained guides.

Amateur geologists occasionally visited the cave, but the first precise survey and map were not made until 1821. And the farther dimensions of the cave system remained unknown until Adolf Schmidl, a 48-year-

A nattily dressed Joseph Nagel points out a feature of a Moravian cave to two members of his expedition. Nagel's scrupulous documentation of his explorations of eastern European caves in the 1740s included drawings such as this one, made by an expedition artist, Carlo Beduzzi.

old professor at Vienna's Imperial Academy of Science, began exploring the Slovenian caves in 1850. Schmidl was both a daring explorer and a formidable intellect, schooled in philosophy and law as well as geology. He ventured along underground rivers in a specially made wooden canoe, which could be dragged through shallow water or even dismantled and carried through dry passages.

On the afternoon of August 30, 1850, when the Pivka River was unusually low, Schmidl and his son launched their tiny craft and entered Adelsberg. For most of the night they paddled upstream, scraping through two low-ceilinged passages that ordinarily were impassable sumps. Meanwhile, an evening thunderstorm had drenched the surrounding area, and at about 1 a.m. the river, swollen by runoff, rapidly rose nine feet, sealing the sumps and stranding the two explorers. For several anxious hours the pair waited in the clammy darkness, 1,800 feet inside the cave, conserving their supply of candles and lamp oil; when the river fell, they quickly departed. Later, aided by a professional surveyor, Schmidl prepared a precise map of the newly discovered river passage and meticulously reported on the subterranean flow of air and water.

But the primary interest in Adelsberg Cave remained commercial. After a railway line was built from Trieste to Adelsberg in 1857, special excursion trains made regular trips to the cave. By 1872 the cave commission was installing its own narrow-gauge railway inside the cavern, although at first

Joseph Nagel uses a pair of geese to help him survey a murky, water-filled cavern in Moravia in 1748. A board rigged with a torch was tied to each bird; when pelted with stones, the frantic geese towed the floating torches in all directions, lighting up the cave.

teams of guides had to push the little two-seat cars themselves, trundling the tourists through the cave like baggage in handcarts.

Some visitors thought the commercialization was overdone. "In the presence of this splendid freak of nature," wrote a *New York Times* correspondent in 1881, "it is somewhat disenchanting to find that the whole thing has been reduced to a system, and that you are confronted by a neat little wooden bureau, where a courteous old gentleman meets you with an inquiry as to 'which kind of illumination you wish to order.' Meanwhile your landlord, with a steadfast eye to business, essays to tempt you by sowing the tablecloth with photographs of the cave and its surroundings."

For most visitors, however, the cave's dazzling variety of stalagmites, stalactites and crystal formations overwhelmed such complaints. And even the *New York Times* correspondent was mesmerized by the glistening stone galleries—"carved battlements, wide-mouthed gargoyles, slender Moorish

In this 1689 engraving, the Austrian explorer Baron Johann Valvasor depicted the stalagmites and dripstone formations of a large chamber inside Adelsberg Cave as hideous gargoyles and fiendish monsters. Valvasor explored about 70 Slovenian caves, portraying each in the same fanciful terms.

columns, grim low-browed arches, fretted roofs, somber Gothic gateways, intertwined spirals, massive pillars festooned with cypress or palm leaves, tomb-like crypts, colossal chandeliers of cold white stone glittering diamond-like with countless drops of water." There is "something indescribably weird and unearthly," he wrote, in "the hollow roar of the Stygian stream below, the ghostly glimmer of its half-seen waters, the mighty void of the sunless cavern which looks all the vaster for those tiny specks of light which struggle in vain against the gloom of this shadow of death."

In America, attitudes toward caves were generally less romantic and more pragmatic. For thousands of years, Indians in the Kentucky karst belt, the most extensive cave region in America, had mined gypsum, a mineral found in cave formations, and used it to make ceremonial paints. Settlers on the rugged western frontier had an even more intense interest in caves: By the late 18th Century they had learned how to extract saltpeter, a vital ingredient of gunpowder, from nitrate-laden cave soil, which had been enriched by the guano from generations of bats.

According to legend, in the 1790s a hunter named John Houchin tracked a wounded bear through the wooded hills of central Kentucky to a large pit near the Green River. After dispatching the animal he entered its lair and found a huge chamber, its walls black with bats and its floor deep in nitrate-rich soil, recognizable by the ammonia stench of guano. A local speculator named Valentine Simons learned of the promising find and, for $80, bought a 200-acre parcel containing the cave entrance.

As war with England loomed in 1812, the threat of a blockade by the British Navy boosted demand for domestic saltpeter. The cave property changed hands several times, its value appreciating sixfold. One owner, boasting that the cave "could supply the whole globe with saltpeter," changed its name to Mammoth Cave. It would become as famous in America as Adelsberg was in Europe.

During the War of 1812 more than 70 slaves mined the cave's chemical wealth, shoveling soil onto oxcarts, hauling it to the cave mouth, dissolving the nitrate in wooden leaching vats and crystallizing it in boilers. The slaves and their owners had little time and less inclination to investigate beyond the nitrate deposits; these lay no more than 1,800 feet from the cave mouth—within easy flying distance of the outside world, where the bats made nocturnal forays for insects. By war's end, the miners had gouged huge holes in the cave deposits and had produced an estimated 400,000 pounds of saltpeter; but soon afterward gunpowder makers found cheaper saltpeter supplies overseas, and the mining operations tapered off.

The miners provided the cave with a curious new attraction, however. While working a nitrate cave about four miles from Mammoth, they had unearthed a mummy of an Indian woman, sitting upright in a stone coffin buried 10 feet below the cave floor amid an assortment of beads, feathers, snake rattles, a deer-foot talisman and utensils. The mummy was nearly six feet tall yet weighed only 14 pounds, apparently because of the desiccating effect of the guano. One of Mammoth's owners, an amateur historian named Charles Wilkins, learned of the discovery and arranged to take custody of the mummy—nicknamed Fawn Hoof for the talisman found in the crypt—and display it in the cave.

Fawn Hoof had been a sort of macabre guest in Mammoth for nearly a year when she was seen in 1815 by an Ohio physician named Nahum Ward, who immediately perceived exciting commercial possibilities. Ward arranged to exhibit the mummy in Lexington, Kentucky, and the shriveled apparition drew such crowds that he went on to travel the Eastern Seaboard, displaying the "Mammoth Cave Mummy" in Philadelphia, Boston and other cities. He also wrote a promotional article, embellished with a map of the cave and an engraving of the departed Indian woman, that was reprinted in dozens of magazines, broadsides and books in the United States and abroad. Before long, Ward's strenuous promotions had earned Mammoth Cave a place in an 1821 volume titled *The Hundred Wonders of the World.* Eventually the mummy was dispatched to the Smithsonian Institution.

Although the tour had made Dr. Ward wealthy, Mammoth still received a mere trickle of visitors. Its fortunes changed, however, after the cave and the land around it were sold in 1838 to Franklin Gorin, a lawyer from nearby Glasgow Junction. The turnabout was due less to Gorin's management than to the special gifts of Stephen Bishop, the 17-year-old slave he introduced as a tour guide.

Athletic, resourceful and fascinated by the countless surprises that the cave yielded, Bishop—always called Stephen—became one of America's most renowned underground explorers. Travel writer Bayard Taylor described him as "a slight, graceful, and very handsome mulatto with perfectly regular and clearly chiselled features, a keen, dark eye, and glossy hair and moustache. He is the model of a guide—quick, daring, enthusiastic, persevering, with a lively appreciation of the wonders he shows." Stephen, Taylor wrote, could discuss geography, history, literature and Greek mythology as well as the geology of his domain. "No one can travel under his guidance," Taylor continued, "without being interested in the man, and associating him in memory with the realm over which he is chief ruler." Such well-publicized encomiums soon swelled the tourist trade.

Dressed in his guide's outfit—a brown slouch hat, jacket and striped pants—Stephen led lamp-toting sightseers through a chamber called the Rotunda, a passage known as the Little Bat Avenue for its winter residents, an 80-foot-high side gallery called the Methodist Church and other fantasies in stone with similarly evocative names.

In his idle hours Stephen began to explore passages branching off from the main route. Penetrating a narrow aperture behind a large flat-topped rock called Giant's Coffin, he edged through a maze of virgin passages and found pieces of a cane torch left behind by Indian gypsum miners centuries before; nearby he found an immense, unfathomable pit. Stephen returned with owner Gorin and another man, who lowered him 90 feet to the bottom of what they christened Gorin's Dome. Such pits eventually led to dripping passages that contained larger, more colorful rock formations than the dry upper cave.

When an adventurous visitor from Kentucky named H. C. Stevenson asked to see an unexplored portion of the cave, Stephen led him through the jagged, twisting passages beneath Giant's Coffin, and on past Gorin's Dome to the Bottomless Pit, a six-foot-wide shaft whose opening entirely blocked the main corridor. Stevenson dropped a rock into the void and counted two and a half seconds before it hit. Then he and Stephen gingerly extended across the abyss a wooden ladder that Stephen had stowed nearby. Holding a flickering oil lamp in one hand, each man in turn straddled the rickety bridge and edged across the abyss. They jumped over a smaller hole, then duck walked down a low-ceilinged oval stoopway to a high, sand-floored passage. There, Stephen let his excited client become the first man to plant his footprints. A few more yards brought them to the brink of a canyon; below, they saw a sizable river—the first major watercourse discovered in Mammoth. Before turning back, they named it the River Styx.

In this way Stephen continually extended the cave frontier, squeezing through tortuous openings barely large enough for his lithe frame and emerging into chambers no man had ever seen. In his first year Stephen doubled the known length of the cave—and, equally important, generated a flurry of newspaper stories about the cave's receding frontier and its intrepid guide.

Unlike the notoriety of Fawn Hoof, the new publicity generated a long-awaited surge of tourism. Suddenly prosperous, Gorin bought two more slave guides for Stephen to train, built a cedar bridge over the Bottomless Pit and introduced a skiff that ferried sightseers along the Styx. Stephen himself became a major attraction, in part because a slave was not expected to be so intelligent and well-spoken. Tales of Stephen's strength and courage became part of the cave's growing lore. He carried one exhausted tourist for six miles and tracked down another who had strolled off alone, lost his lamp, knocked his head on the rocks and lain unconscious for 43 hours.

Within the year the flood of tourists overwhelmed the small log structure that Gorin used as an inn. Lacking the capital to expand, he sold the whole enterprise—cave, land, slaves and inn—to a wealthy Louisville doctor and gentleman-farmer named John Croghan in the fall of 1839.

Working for this new master, Stephen made yet another splendid discovery, seemingly on a whim, by climbing a slope above the river and following a long alley to a boulder-strewn choke. He and a visitor clawed away rocks with their hands, not knowing whether they would find a solid rock wall or another chamber. After hours of work they reached an opening and wriggled through, emerging on a ledge midway up the sloping wall of the deepest pit Stephen had yet seen. As they cautiously descended, their lamps illuminated rock ornaments rivaling those at Adelsberg. The ceiling of this great shaft, soon named Mammoth Dome, was invisible in the blackness 192 feet above the floor. In subsequent months Stephen explored a second stream, the Echo River, and opened up several more miles of broad boulevards, one of them leading to an enchanting chamber festooned with gypsum flowers and snowball-like growths on the walls—the Snowball Room.

To publicize Stephen's discoveries, which had long since outdated the 1835 map, Croghan resolved to print new maps and a lavish guidebook,

This engraving is one of the few pictures of the legendary slave guide of Mammoth Cave, Stephen Bishop. When Bishop died in 1857 at the age of 36, his talents as an explorer and self-taught geologist had already made him almost as famous as the cave itself.

A 19th Century engraving shows a party of tourists being paddled along Echo River, one of the largest underground streams in Mammoth Cave. While one stocking-capped guide regales an enthralled female passenger, another blows a horn to demonstrate the cave's wondrous echo effects.

"bound in neat half morocco with cloth sides, lettered and filletted on the back," according to the printer's contract. He set Stephen to work drawing a pencil map from memory during January 1842. Within a week the uneducated slave had produced an elaborate freehand sketch showing the cave's hundreds of sinuous passages, all drawn to scale—a document whose usefulness endured for 40 years.

Croghan enlarged the cave hotel from a two-room log cabin into a luxurious complex consisting of four main buildings and 16 cottages. He also persuaded the state legislature to help rebuild the wretched, nearly impassable roads between the cave and the main Louisville-Nashville stage route. Soon travelers were paying more for accommodations and meals than for the various cave tours, which ranged in length from two to nine miles. But Dr. Croghan had another, purely philanthropic project in mind. For years he had believed that the nitrates, pure air and constant temperature within caves might cure consumption, the 19th Century scourge that is now called tuberculosis. Several other physicians shared this conviction, founding their belief on reports that saltpeter miners had enjoyed unusually good health in the sunless atmosphere, and that a walk through such a cave, whatever the distance, caused surprisingly little fatigue.

Croghan decided to test the thesis. In 1842 he built a dozen cottages about a quarter mile inside the cave and accepted his first patient, a medical man who had diagnosed his own ailment as pulmonary consumption. After five weeks the physician departed, pronouncing himself "very much relieved." News of the apparent cure prompted several other doctors to refer patients to Croghan, but the initial enthusiasm quickly waned. Several patients left after short stays produced no improvement in their health, and two men died in the cave, withering like the shrubs the patients had planted in a pathetic attempt to embroider their bleak surroundings. Within a year the experiment was abandoned entirely. More than a decade later, Bayard Taylor shuddered at the sight of the would-be sanatorium: "The idea of a company of lank, cadaverous invalids wandering about in the awful gloom and silence, broken only by their hollow coughs—doubly hollow and sepulchral there—is terrible."

Mammoth Cave and its celebrated guide continued to thrive as a tourist attraction, however, its popularity eventually rivaling Niagara Falls. By

1850 the cave boasted 226 avenues, 47 domes, 23 pits and 8 waterfalls, most of them easily accessible from paths blasted into the rock by Croghan's workmen. The cave hotel became an exclusive summer resort for wealthy Louisville families and a standard stop for visitors to the area. The renowned Norwegian violinist Ole Bull gave a recital inside the cave, in a room known from then on as Ole Bull's Concert Hall; soprano Jenny Lind, the "Swedish Nightingale," visited a few years later.

In 1849 Croghan died—ironically, of consumption. His will left the cave to his family and stipulated that Stephen be freed seven years later. Once free, Stephen planned to buy the freedom of his wife and son and move to the recently established African republic of Liberia, but his dream was never realized: He died in 1857 at the age of 36, one year after becoming a free man. Years later Franklin Gorin, Stephen's former master, eulogized him aptly: "Stephen was a self-educated man. He had a fine genius, a great fund of wit and humor, some little knowledge of Latin and Greek and much knowledge of geology, but his great talent was a knowledge of man."

The Civil War put a stop to tourism, and business was slow to recover in the postwar period. Mammoth's owners worked hard—if not always ethically—to stimulate the trade. In 1875 another female mummy, discovered at nearby Salts Cave, was advertised as the original Fawn Hoof and exhibited in Mammoth. And Mammoth guides continued to arrange crowd-pleasing shows for visitors, good-naturedly tweaking the tourists' uneasiness. "A nervous person," a sightseer from New York wrote, "imagines great eyes staring at him from every corner, and cold hands, ready to grasp his shoulder, at every step."

Mammoth no longer had the field to itself. Rival showplaces like Wyandotte Cave in Indiana were also doing good business, in part because the nation's growing network of railroads made travel so much easier. In the eastern United States, the most resounding success was experienced—for a time—by three denizens of Virginia's Shenandoah Valley: Benton Stebbins, a struggling, 53-year-old photographer; Andrew Campbell, a 42-year-old tinsmith; and Billy Campbell, Andrew's 26-year-old nephew. Knowing that a railroad would be built across the valley, and that there were several small caves nearby, the opportunistic trio spent so much time

Tourists in Kentucky's Mammoth Cave share a novel underground picnic with their two black guides in 1892. By this time, the cave's facilities had deteriorated so much that tourism had fallen sharply from its pre-Civil War peak of 40,000 visitors a year.

A pile of flat stones emplaced by visitors from Kentucky reaches Mammoth Cave's 12-foot-high ceiling. Tourists were encouraged to leave such evidence of their visits until the National Park Service forbade the practice in 1941.

running down cave rumors and probing the valley's many shallow pits that they became known locally as "phantom chasers."

On the steamy morning of August 13, 1878, as they stopped to rest in a clump of brush on the side of sinkhole-pocked Cave Hill, an aptly named ridge about a mile from the town of Luray, they stumbled on what seemed to be a large rabbit hole at the bottom of a 10-foot-deep sink. As they slid into the sink they felt a cool breeze whispering up from the hole. After fetching tools and rigging a block and tackle to lift dirt and rocks out of the sink, the men labored with picks and shovels for days, enlarging the opening until Andrew Campbell, the smallest of the three, could slip through. Holding on to a rope, he slid down a muddy slope, lighted a candle and found himself in a narrow rift about 15 feet long and five feet wide, with a small hole at one end. He let go of the rope, squeezed through the hole, and stood rapt in a large, glittering, circular chamber about 10 feet across and perhaps 40 feet high. A slender white column in front of the hole reached from the floor to the ceiling, while, overhead, hundreds of sword-shaped stalactites reflected the light. A worried shout from his nephew Billy summoned him back to the surface.

Stebbins and the Campbells returned that night to explore further. The cave seemed to have an unlimited supply of wonders. One passage wound through massive stalagmites and delicate stalactites for several hundred yards; another led to a chamber roofed with soda-straw formations, which glistened bright red and white and tan in the candle glow. Andrew Campbell wandered along a broad, lofty boulevard, then stepped into a pool of water that filled the passage. He waded on in cold, waist-deep water, but began sinking in mud and slogged back out. His candle flame was still being blown back toward the cave's entrance, a sign of extensive passages ahead that could only be explored by boat.

Having found a splendid cave, the explorers now needed either to strike a bargain with the landowner who controlled the entrance or, better still, to buy the parcel themselves. Secrecy was essential. After emerging from the cave, they blocked its entrance with rocks, then changed their muddy clothes in a nearby field to hide the evidence of their nocturnal foray. The next day Billy Campbell quietly perused the courthouse records. The 28-acre parcel, once owned by a bankrupt merchant, had recently been sold at

Spectators throng Giant's Hall in Luray Caverns in Virginia during an 1878 "illumination"— a promotion that involved lighting the cave with hundreds of candles. For an additional fee, guests could dance on a specially laid wooden floor to the music of an orchestra.

39

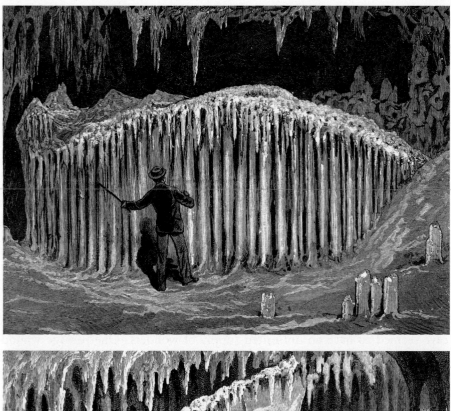

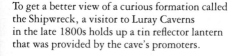

A guide taps out a tune on 56 graduated columns of dripstone in Luray Caverns in Virginia. The formation, dubbed the Organ, was a popular stop on the early tours of the cave.

To get a better view of a curious formation called the Shipwreck, a visitor to Luray Caverns in the late 1800s holds up a tin reflector lantern that was provided by the cave's promoters.

auction for eight dollars an acre, but the buyer had so far failed to make the required $28 deposit. If he did not consummate the purchase, it would be auctioned off again at the beginning of September. This time it might go for as much as $20 an acre and require a $120 down payment—an enormous sum for three impecunious tradesmen.

During the next month the trio secretly raised $40 each. Stebbins obtained a loan by using as collateral his wife's organ, a prized possession that she had inherited from her parents; Andrew Campbell swore a Masonic brother to secrecy and asked him for a loan; Billy Campbell borrowed from his father, the Page County Sheriff. Then they waited to see whether the previous buyer would put down the deposit, all the while pretending to search for caves and never visiting their discovery.

The property went on the block on September 10. To conceal their interest, the explorers had Billy's father, a courthouse regular, represent them. Bidding was surprisingly hot. The bankrupt's family were buying up his estate, and their surrogate drove the price of this rocky, seemingly worthless tract to $17 an acre before yielding. The new owners immediately paid the $96.90 deposit, recorded the sale and began to celebrate. They were in business—or so they thought.

As they began to plan their show cave, the new owners worried that its chambers, while grand, might be too small to attract tourists. In mid-September they returned with a dinghy, enlarged the entrance to accommodate it and launched their little craft on the lake. They drifted soundlessly down the broad, watery avenue, passing under a sculpted rock bridge and nosing into shore at the base of a muddy slope. At its top, a long, level corridor led to a vast chamber walled with parti-colored flowstone draperies. Enormous columns and stalagmites met them at every turn; folds of orange and red curtained the walls. They clambered about for two hours in overjoyed awe before returning to the boat.

Stebbins and the Campbells fell to work readying the cavern for visitors—building stairways, bridges and handrails—and tourists began arriving almost immediately. An enthusiastic *New York Herald* account of the cave's discovery attracted visitors from the entire Eastern Seaboard and furnished a simple, catchy name—Luray Caverns. In November, Stebbins, a natural showman, placed thousands of candles throughout the cave and invited the public to a "grand illumination" at 50 cents per adult and 25 cents for children, netting $91 profit in one day. The overnight success of the phantom chasers prompted hordes of men to scour the countryside in search of comparable bonanzas. "There is not a promising rat hole in Page County," the *New York Herald's* man observed, "that has not been 'opened up' in the hope that it might lead to great cavernous spaces below."

But the good fortune of Stebbins and the two Campbells was short-lived. Early in 1879, relatives of the former owner of the cave property, seeking to save the rest of his far-flung estate from the auction block, filed a lawsuit charging that the group had defrauded them by concealing knowledge of the cave at the original sale. Until the suit was settled, no bank would loan money for desperately needed improvements—boats, bridges, footpaths and the like. As the case dragged through the courts, Stebbins arranged to sell the property to the Shenandoah Valley Railroad Company for $40,000, enough to make all three partners rich for life—but they could not consummate the sale until they won the lawsuit.

The circuit court found in favor of the three explorers, but on April 21, 1881, the Virginia Supreme Court unanimously ruled against them. The original owner's family reclaimed the cave, then turned around and sold it to the railroad for $39,400. Under its new operators, Luray Caverns would become one of the best-known and most profitable caves in America. Stebbins and the Campbells got nothing, although Billy Campbell was subsequently employed as a cave guide.

Just before the hapless threesome lost the cave, they welcomed a scholarly delegation from the Smithsonian Institution to Luray. The report of this august panel was a marvel of fevered imagination that neatly illustrated the popular attitude toward caves in 1880: "Here in this dark studio of nature are reproductions of all those objects which fill the mind with pleasure, wonder or alarm," among them "spectral beings—terrestrial, celestial and infernal." The prose was strikingly similar to the contemporary *New York Times* article about Adelsberg Cave, which dwelled on "the hollow roar of the Stygian stream below, the ghostly glimmer of its half-seen waters," and "the gloom of this shadow of death." Such romantic accounts did not consider how these caves had come into being, or what processes accounted for their spectacular formations and exotic inhabitants. Such questions—and their answers—awaited the birth of a new kind of science in the next century. **Ω**

The first cavern to achieve worldwide renown as a showplace was Adelsberg Cave, located near a village of the same name in Austria. Explorers began probing it as early as the 13th Century, but the cave received only modest attention until the Holy Roman Emperor Francis I decided in 1818 to make a firsthand inspection of its reputed wonders.

The royal visit was preceded by a flurry of activity in the cave as workers cleared away rubble and climbed rickety ladders to install torches. In the process, one laborer happened upon a gap in the cave's far wall, some 90 feet above the floor. Crawling through the opening, he found a new section far larger than the first, bedecked with fantastic dripstone formations.

The Emperor did not venture into this portion of Adelsberg, but local officials reasoned that, if problems of access could be conquered, the public would find it irresistible. At their behest, workers leveled paths through the chambers, built a wooden bridge across the cave's underground river and chiseled stone stairways into the walls *(overleaf)*. The crude torches were replaced with chandeliers and oil lamps, and guides were trained. To illustrate a deluxe guidebook, an artist named Alois Schaffenrath executed 19 watercolors of the cave; a sampling appears here and on the following pages.

Adelsberg was soon drawing 1,000 visitors a year—and that figure was quickly surpassed after a railway line was opened to Adelsberg in 1857. By the 1870s, the annual number of tourists had soared to 8,000 and a fashionable hotel had been built nearby. Even Americans, amply endowed with caves of their own, were forced to acknowledge the appeal of this European gem.

After visiting Adelsberg in 1881, a reporter for *The New York Times* wrote in awe of the "stalactites of unequaled splendor" and the "fantastic architecture" in the cave. Indeed, concluded the *Times* man, although Adelsberg lacks "the mighty expanse of the Mammoth Cave of Kentucky, every part of it is filled with a stern and gloomy grandeur which is indescribably impressive."

In a watercolor of Adelsberg's exterior by Alois Schaffenrath, tourists line up to enter the cave by way of a hillside portal. Below, the Pivka River emerges from another opening.

Inside Adelsberg's Great Hall, visitors pass
between two upper galleries by descending a
staircase, crossing a bridge over the Pivka
River and climbing a second staircase on the
opposite side. A chandelier, oil lamps
and the guides' torches light up the dripstone
formations of the 115-foot-high vault.

Three tourists admire a nearly transparent stone drapery known as the Curtain. As in many of his cave views, the artist whimsically depicted himself sketching the scene.

Their way barred by fallen columns, tourists prepare to leave Adelsberg Cave on a well-constructed path. In this painting, Schaffenrath included not only himself *(lower left)* but three assistants who are illuminating the far reaches of the chamber and his sketch pad.

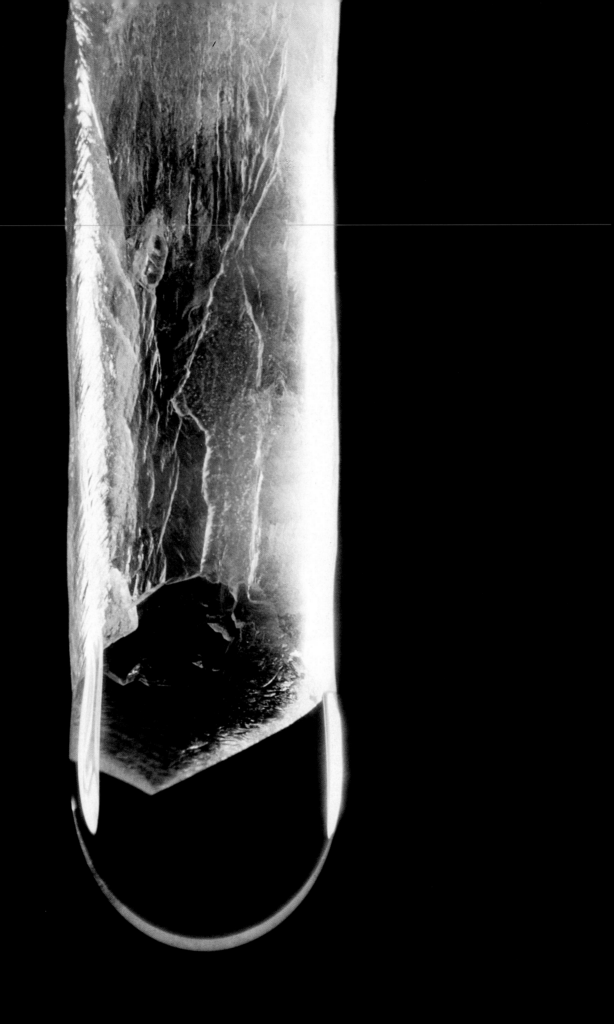

THE BIRTH OF SPELEOLOGY

Early cave adventurers were a hardy, quixotic breed who ventured underground to emblazon their names in record books or to make easy fortunes. Science, as such, did not enter into their haphazard quests. The cavers employed rudimentary equipment—rope and wooden ladders, windlasses and candles—that had not changed much since John Beaumont's 17th Century explorations, and their impressionistic reports on the subterranean world contained little factual information. Not until the final decades of the 19th Century did underground expeditions gain scientific respectability, and when they did, it was chiefly through the methodical work of a young Parisian, Édouard-Alfred Martel.

Even as a child, Martel, by his own account, was "a born troglodyte, as fond of burrowing under ground as a badger." In 1866 his father, a lawyer and amateur paleontologist, took the precocious seven-year-old to the Grottoes of Han, a famous show cave in the Belgian Ardennes. While touring the cave's lofty balconies, Martel suddenly recognized his lifelong calling. "As I stood awe-struck, gazing down into the very bosom of the mountain, on the inky, torch-lit surface of a subterranean lake," he later wrote, "I made up my boyish mind that I would devote my life to exploring those wonderful recesses in the earth." During summer vacations thereafter he visited caves in the Pyrenees, Switzerland and Italy, and became an accomplished alpine mountain climber, which helped him greatly in caving. In school he was a keen student of geography, graduating with first prize in the 1877 national examination. Two years of compulsory Army service interrupted Martel's studies, but at its conclusion he immediately renewed his subterranean education with a visit to Adelsberg, the jewel of European caverns.

However, in 19th Century France such unsalaried pursuits as exploring caves were reserved for wealthy gentlemen of leisure. Martel had pledged to follow his father into the law—a promise that he later regretted. After he completed his legal studies in 1883, Martel began devoting all of his vacation time to cave expeditions into the Causses of southwestern France—arid limestone plateaus that he described as "naked deserts, gloomy, monotonous, without water or wood, and almost without inhabitants." At first the untutored explorer cut a somewhat comical figure, descending into caves wearing a lawyer's high-collared shirt and bowler hat. But during a series of small expeditions, financed by his meager savings and conducted with a handful of friends and helpers, Martel rapidly developed a disciplined and methodical system that would transform cave exploration.

Suspended by surface tension, a drop of water bulges from the bottom of a soda-straw formation in Beauty Cave, Arkansas. Soda straws, often the precursors of stalactites, are created when water drips slowly from a cave ceiling, depositing calcite in a tiny circle around the upper rim of the droplet.

After five years of experimentation, Martel in 1888 began a series of annual "campaigns," as he called them. During his vacation each summer, he set off with several friends, two trusted foremen and two bullock wagons full of equipment—pulleys, windlasses, ladders, lamps, silk and hemp ropes, hatchets, provisions, photographic apparatus, portable beds and assorted baggage. At first glance local inhabitants sometimes took this motley caravan for a traveling circus. When they learned of his true purpose, the superstitious rustics usually greeted "the gentleman who traveled into holes" with considerable trepidation, perhaps believing that he was tempting the devil. Martel often had trouble hiring 10 stalwart peasants to manage his equipment, even at seven francs a day. "The old women would cross themselves," he reported, "and mumble between paternosters: 'You may go down if you like, but you will surely never come back.' "

At the cave's mouth (usually a pothole, the type that predominates in much of Europe) Martel's men first drove crowbars into the ground around the chasm and strung a rope barrier from them; this was to hold back the curious who otherwise would crowd to the very edge, loosening dirt and stones, which would rain down on the explorers below. To determine the length of rope ladder required, Martel would plumb the pothole with a six-pound cannonball tied to a heavy line. (Sometimes the weight came to rest on a towering pyramid of stones at the bottom of the shaft and Martel

French lawyer Édouard-Alfred Martel reflects the dignity of his profession in this portrait taken in the 1920s. His avocation—the scientific study of Europe's great caverns—earned him world fame in the late 19th Century as the founder of modern speleology.

would be left dangling on too short a ladder when he descended.) Then his helpers would secure a rope and pulley to a roughhewn beam or, at large pits, a wooden tripod rigged over the funnel-shaped cave mouth.

While his men arranged the ropes and winches, Martel would don specially made coveralls of his own design, with a multiplicity of pockets for a whistle, six large candles and several stubs, magnesium wire (for brighter light) with a wooden holder, matches, flints and steel, a hammer, two knives, a plumb line, a 10-yard tape measure, two thermometers, two barometers, pencils, a compass, a notebook, a first-aid kit, several cakes of chocolate and a flask of rum. He would then strap on his proudest innovation—a 14-ounce portable telephone. During his early explorations Martel had discovered that in deep holes shouts and even whistle blasts were inaudible at the surface, so he adapted the French Army's battery-powered field telephone to caving. Although he frequently telephoned simply to arrange some flourish of Gallic panache, such as instructing his helpers to lower an extra bottle of wine, he thought of the apparatus as a vital safety precaution, "doubling our boldness by the assurance of help in case of need."

During every descent Martel was a stickler for safety, wearing a life line in case he fell from the ladder. Wiry and nimble, he preferred to use a rope ladder even for deep pits, but in bell-shaped potholes the dangling, unsupported ladder was nearly impossible to climb. In such situations Martel would sit on a sturdy two-foot plank, a sort of bosun's chair, while helpers lowered him with a half-inch rope. He hated such descents because the unwinding rope would twist and spin him around until he was dizzy and nauseous. "The only way to keep from losing one's head," he advised, "is to count the turns."

Even such meticulous precautions could not entirely avert danger. Once while he was twirling at the end of a 400-foot rope, Martel smelled something burning and noticed that his head felt quite warm; a carelessly strapped-on candle had ignited his felt hat, and the flame threatened to burn the rope. As he dangled, Martel doffed the cap, knocked out the fire—and carefully preserved the charred headpiece as a memento.

Odorless carbon dioxide gas, which sometimes accumulates in unventilated crevices, posed a subtler hazard. As Martel descended a rope ladder into one French cave, he felt himself growing giddy from lack of oxygen but foolishly continued while holding his breath—and rapidly lapsed into unconsciousness. He fell from the ladder but was saved by alert crewmen aboveground who hauled their nearly asphyxiated leader to the surface with his safety line and revived him. On another occasion Martel's best foreman, a blacksmith named Louis Armand, became stuck in the tiny exit from a chamber, not only trapping Martel and his cousin Gabriel Gaupillat but depriving them of fresh air. "Our candles soon sputtered out and we began to gasp and choke," Martel recalled. "At last with a supreme effort the burly cork was got out and we breathed again."

Martel's innovations sometimes created new dangers. Instead of the rigid, dismantled canoe that Adolf Schmidl had dragged into Adelsberg Cave in 1850, Martel employed American-made folding canvas canoes, christened with names such as *Alligator, Crocodile* and *Caiman*. With these 40-pound craft, which were lowered in two compact bags and assembled underground, Martel traversed daunting subterranean rivers and lakes. On one occasion he and two companions were crossing a lake by candlelight when they bumped into a low roof they had failed to see. All three were knocked into the chilly water, which of course extinguished their lights. Martel's companions luckily soon found the shore, but Martel himself was disoriented in the darkness and had no idea in which direction to swim. As his sodden clothes began to drag him down, the capsized boat suddenly bobbed to the surface directly beneath him and he climbed astride—only to be knocked back into the water by the same low roof. Exhausted, he paddled for his life, shouting, "Help! I'm drowning!" until his companions repeatedly called out to give him his bearings. "I must have traveled halfway round the lake," he recalled, "for their welcome cry reached me from a considerable distance."

When Martel finally scrambled ashore, the three shivering men found that while they had extra candles, the matches in their pockets were soaked. Now another of Martel's precautions paid off. He had been carrying two metal boxes of matches wrapped in oilcloth in the breast pockets of his flannel shirt. One box was full of water, but the second was perfectly dry

Deep within Padirac Cave in France in 1890, Martel and his companions leave their portable boats and use a ladder to reach an opening above an underground lake. Beyond this aperture Martel found a smaller, elevated lake.

and quickly yielded what Martel thankfully called "the spark of deliverance." With the candles lit, he reported, "A first look round showed that we had nothing to fear. The boat was lying under the projecting roof; we had only to swim across and recover it."

Martel engineered vast improvements in both the reliability and reach of the caver's ropework—ladders, windlasses, pulleys and the like—enabling him to conquer deep vertical pits; in 1889, he descended into the great pit of Rabanel, about 100 miles northwest of Marseilles, and did not touch bottom until he had gone down 700 feet, several times the previous record. But the elaborate rigging necessary for such feats involved considerable risks. Martel once dangled for 45 minutes on a rope 260 feet underground, spinning in absolute darkness, while his crew wrestled with tangled lines.

Faulty rigging combined with careless procedure nearly killed Martel's second foreman, Émile Foulquier. On this occasion Foulquier was the last man to leave a 200-foot pit, after the telephone had been routinely removed. As the crew was heaving on the windlass, the men heard Foulquier bellowing at them; in alarm, they cranked faster until finally they could make out his shouts: "You are crushing me." A rope tied around the poor fellow's chest with a running knot had been squeezing the life out of him. His frantic comrades got Foulquier to the surface just as he fainted. "A cordial gave him speedy relief," Martel wrote, "but that evening my squad had a long object-lesson on the proper method of tying knots." And also, one might hope, on the need to leave the telephone until the last man emerged.

The greatest danger of all was posed by objects hurtling down a shaft. Martel tried to throw down loose rocks as he first descended. But despite all precautions, he could not guard against showers of stones dislodged by dogs or by spectators who ducked under the rope barriers surrounding the cave pit. Dropped equipment was equally dangerous. As Martel's team was hoisting up packages after an exploration of Rabanel, a poorly fastened parcel opened as it reached the lip, raining hammers, drills, canteens, camera equipment and lanterns on the startled explorers, who darted beneath an overhang. "The smallest of these objects would have cracked our skulls," Martel observed. After Martel's companions had preceded him into another pit, a ladder rung knocked a safety lamp from his chest as he was descending, and sent it plummeting as he shouted a warning to men on a ledge below. The lamp missed foreman Armand by inches, knocking a candle from his hand and grazing his foot, then plunged on into the abyss.

Once safely on the floor of a pothole, Martel would begin recording observations in a notebook. To calculate the pit's depth, he would measure the barometric pressure at the bottom and compare it with the surface pressure. He would measure the horizontal dimensions of each chamber with a metal tape, drawing a rough sketch of the cave as he worked. An ingenious device allowed him to make vertical measurements: After tying a length of light silk twine to a small paper balloon, he would suspend a sponge beneath the apparatus, pour alcohol on the sponge and light it. Air heated by the burning alcohol would quickly lift the balloon and twine to the roof, allowing him to determine the height simply by measuring the length of twine. Martel also made a point of measuring subterranean air and water temperatures at several locations, in the process proving that—contrary to previous belief—they varied somewhat both according to season and within a single cave.

Of Martel's many campaigns, perhaps his favorite was the 1889 expedition to the Causse de Gramat, a cave-packed karst plateau between the Dordogne and Lot Rivers. When he inquired about the existence of any

Most of the world's caves were formed over hundreds of thousands of years by the slow dissolution of limestone, but some spectacular caverns are the result of processes that can be completed in a matter of days or weeks.

The lava that spews from a volcano sometimes cools and solidifies rapidly on the surface while remaining fluid underneath. Subsurface streams of molten rock may then course through slowly hardening channels in the lava flow and drain away, leaving a network of tunnels behind. The largest known lava cave, Kazimura Cave in Hawaii, extends for more than seven miles.

Glacial ice is another medium for cave making: Hollowed out by meltwater and warm winds, glacier caves are perhaps the most beautiful of all. The dense ice filters the light into brilliant shades of blue, and the wildly sculpted walls reflect color like the facets of a gemstone. But these caves are also the most dangerous: Melting ice is inherently unstable; roof collapses are frequent; sudden floods of icy meltwater are common.

In arid climates, dust particles driven by powerful winds scour vulnerable strata in cliff walls and create large, shallow caves. Often perched high above the surrounding terrain, wind-carved caves provided safe and enduring shelter for ancient communities in the Upper Nile Valley and in the American Southwest.

Less suitable for permanent human habitation are sea caves, coastal caverns excavated by ceaseless wave action. Over the millennia, waves blasting into weaknesses in rock formations have gouged out thousands of spacious caverns along the seashores of the world.

The blackened, rugged walls of two cave passages in the Lava Bed National Monument in California commemorate an ancient stream of molten rock that diverged into two channels. Located near a shield volcano called Medicine Lake Highlands, these tunnels are part of a lava cave network that underlies 72 square miles.

Pounded by the waters of the southern Indian Ocean, sea caves perforate the shoreline at Port Davey, Tasmania. Some of these caves extend as much as 200 feet inland.

Dressed for a cold, wet tour, a visitor inspects a subglacial waterfall inside a deteriorating remnant of Muir Glacier in Alaska. Because of rapid melting, glacier caves frequently change their shape from week to week.

Safely ensconced in the side of a sandstone cliff, the ruins of an Indian village—now called Cliff Palace—fill a cave carved by water and wind at Mesa Verde in Colorado. This village was built in the 13th Century.

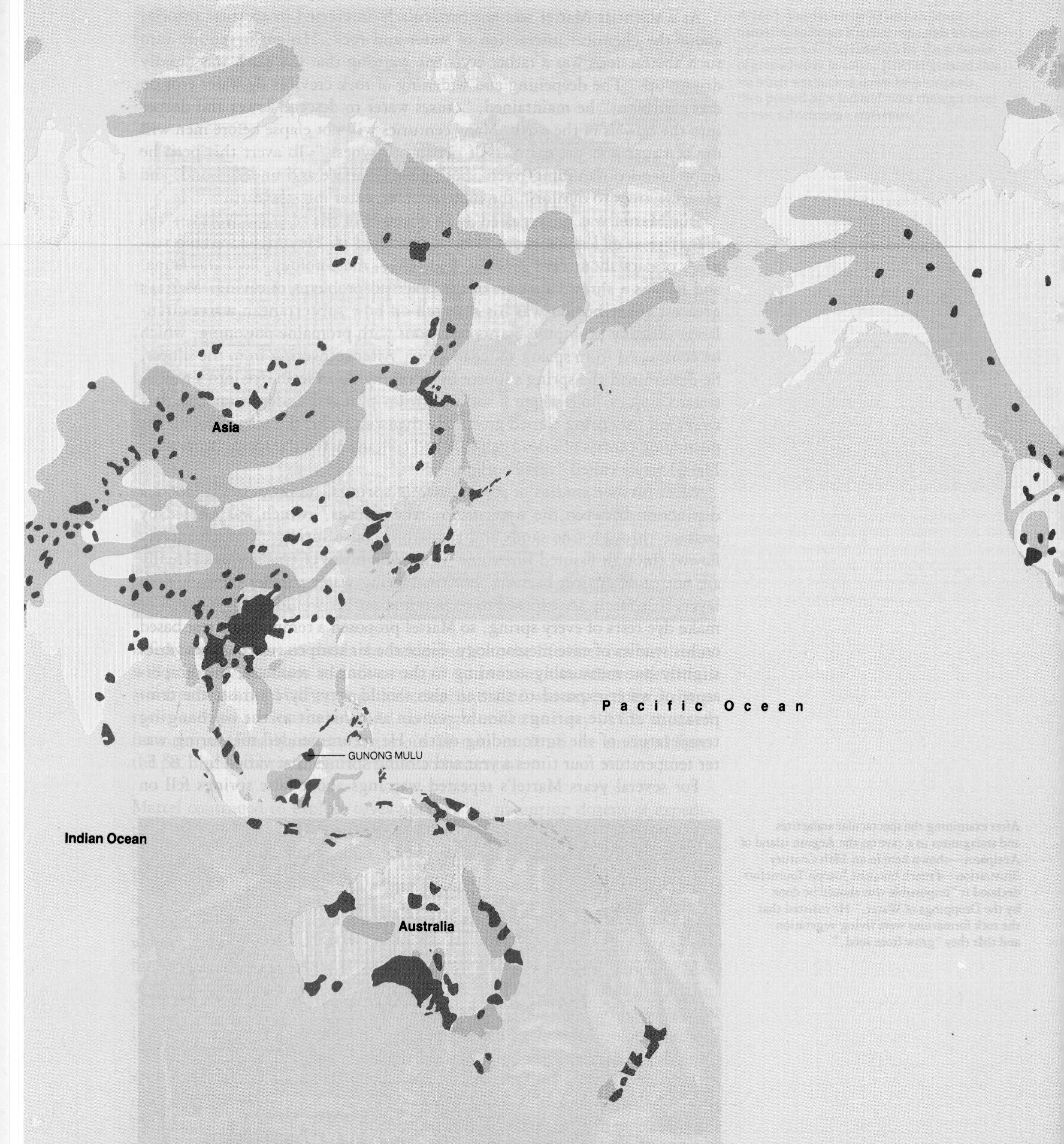

Asia

Pacific Ocean

—— GUNONG MULU

Indian Ocean

Australia

A map of the world's karst regions (*purple*)
—rugged landscapes of heavily eroded
limestone—shows where most caves are found.
The karst is found principally in fold areas
(*blue*), where the movements of continents have
thrust slabs of buried limestone to the surface.

North America

GAPING GILL

HOLLOCH   ADELSBERG

Europe

FLINT RIDGE-MAMMOTH

MONTESPAN

CARLSBAD

PIERRE ST.-MARTIN

EISRIESENWELT

SAN AGUSTÍN

Atlantic Ocean

Africa

South America

Indian Ocean

deaf ears. But when, in 1898, contaminated water caused a typhoid epidemic in a large Army garrison, the ministers of war and the interior immediately arranged conferences with the eminent speleologist. Soon afterward the government required that all sources of drinking water be checked by a geologist, a chemist and a microbiologist. In 1902 a new law banned the dumping of refuse and dead animals near chasms, potholes and underground streams. Martel continually agitated for stricter enforcement of the laws and collected evidence of hundreds of violations, but he also noted with pride that by 1908 the number of deaths from typhoid had been reduced by half. For this tireless battle, the French government in 1923 awarded him its gold medal for epidemics, reserved for those who imperiled their lives in the fight against disease.

By the time Martel died in 1938, at the age of 78, he was acclaimed worldwide as both an explorer and a scientist. The grand old man of speleology had personally probed nearly 1,500 caves, hundreds of which had never been entered before, and his technical innovations had become standard equipment for other cavers. And as the foremost evangelist for the infant science, he inspired a new generation of disciples who would continue the work of scientific exploration.

Martel is honored as the founder of speleology more for his persistence and dedication than for his limited scientific achievements. Speleology always has been a somewhat eccentric discipline, as much sport as science, and its early practitioners were at best descriptive scientists who merely recorded underground observations. Theoretical insights about how caves form came from scientists in other fields—primarily chemistry and geology—who until the end of the 18th Century were themselves limited by faulty assumptions. Early chemical theory was dominated by the idea of phlogiston, a physical substance (rather like an element in modern chemistry) that combustion supposedly released as flame. Principles based on atoms and molecules were unknown. Geology was ruled by a literal reading of the Book of Genesis, which asserted that the earth was shaped by sudden, catastrophic acts of God rather than by gradual processes such as erosion.

Catastrophism, as it came to be called, led philosophers to attribute caves to every conceivable cataclysm—underground fires, explosions, volcanoes and earthquakes—but most early theorists blamed the Biblical Flood. This

Two hikers pause to examine a barren alpine karst area in Wyoming. The limestone, cracked by alternate freezing and thawing, is vulnerable to snow runoff and rain, which eat away at the crevices and the underlying strata.

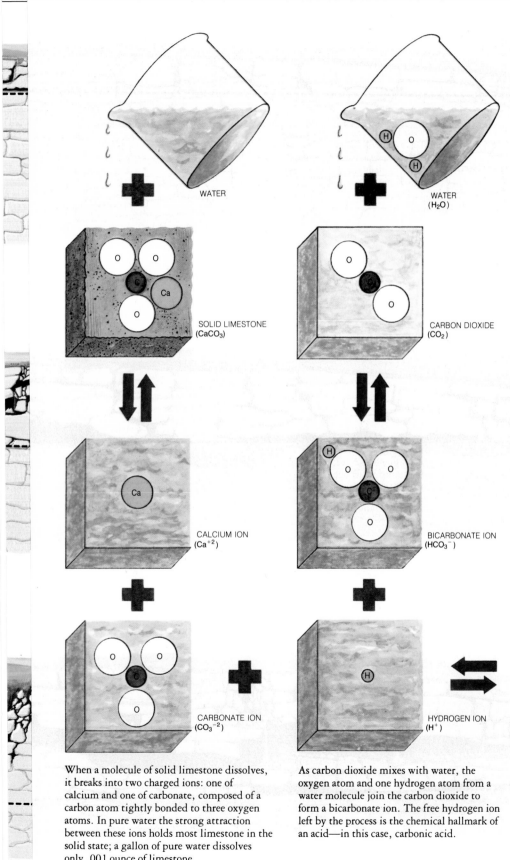

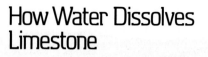

How Water Dissolves Limestone

Caves are carved by water; but acting by itself, in the process diagrammed at far left, water cannot dissolve enough of limestone's main component, calcium carbonate, to create caves. The key to cave building is a chemical reaction that occurs as groundwater seeps downward, through decaying plant and animal matter, dissolving carbon dioxide gas released by the decomposition. As illustrated at near left, some of the gas reacts with water to form dilute carbonic acid.

Independently, both processes—the dissolution of limestone and the formation of carbonic acid—would virtually halt when they reached a state of equilibrium. But one product from each process—a carbonate ion and a hydrogen ion—combine in a third reaction (*bottom row*). This third reaction enables the original limestone reaction to dissolve much more limestone—up to $\frac{1}{10}$ inch per year in some caves.

When limestone-laden water begins dripping through a cavern, however, the chemical reactions are reversed. Much of the carbon dioxide bubbles away into the subterranean atmosphere. Thus weakened, the carbonic acid rapidly becomes saturated with dissolved ions and begins to drive the limestone dissolution backward, forcing calcium carbonate to gradually precipitate from the solution in sparkling subterranean stalactites and stalagmites.

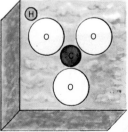

WATER

WATER
(H_2O)

SOLID LIMESTONE
($CaCO_3$)

CARBON DIOXIDE
(CO_2)

CALCIUM ION
(Ca^{+2})

BICARBONATE ION
(HCO_3^-)

CARBONATE ION
(CO_3^{-2})

HYDROGEN ION
(H^+)

BICARBONATE ION
(HCO_3^-)

When a molecule of solid limestone dissolves, it breaks into two charged ions: one of calcium and one of carbonate, composed of a carbon atom tightly bonded to three oxygen atoms. In pure water the strong attraction between these ions holds most limestone in the solid state; a gallon of pure water dissolves only .001 ounce of limestone.

As carbon dioxide mixes with water, the oxygen atom and one hydrogen atom from a water molecule join the carbon dioxide to form a bicarbonate ion. The free hydrogen ion left by the process is the chemical hallmark of an acid—in this case, carbonic acid.

When carbonic acid dissolves limestone, carbonate ions and hydrogen ions from the two vertical reactions at left combine (*bottom row*) to form a second bicarbonate ion, thus creating a momentary shortage of carbonate ions at far left. To restore equilibrium, extra limestone dissolves—perhaps 25 times the amount that would dissolve in pure water.

A PROFUSION OF WATER-WORKED WONDERS

Water, the shaper of caves, is a versatile as well as patient worker; even as it completes the excavation of a gigantic chamber in the earth, it is attending to the infinitely delicate task of decoration. Setting microscopic crystals of mineral one upon another, the slowly moving liquid crafts an eerie gallery of sculpture whose variety and beauty can be breathtaking.

The individual formations are called speleothems—a term derived from the Greek words for "cave," *spelaion*, and "deposit," *thema*—and almost all of them are made of crystals of calcium carbonate in the form of calcite *(page 67)*. Their multitudinous shapes, from the narrowest, most fragile soda straw to the broadest, most mountainous stalagmite, are dictated by the water's various movements.

Where water drips incessantly into a cave, conical stalactites descend from the ceiling and rounded stalagmites rise from the floor. In time, a stalactite and stalagmite may meet and grow together into a massive floor-to-ceiling column *(right)*. Water that trickles along a slanted ceiling deposits delicate sheets called cave draperies. When water seeps slowly through bedrock, emerging into the cave from minute fissures, it can build wildly contorted structures known as helictites. Each helictite grows as water is forced up its tiny central tube by capillary action, forming crystals at its tip *(page 85)*.

Still other speleothems grow in or on cave pools. Round, thin patches of calcite called cave rafts may form on the surface of an undisturbed pool, and thicker deposits of shelfstone may extend from its sides. Water that gently laps over the edge can leave a long rimstone dam along its water line. And in shallow pools fed by dripping water are found perhaps the most exotic of all speleothems, cave pearls. These small spheres, produced by the accretion of calcite layers around grains of sand, bear a remarkable resemblance to the precious sea-born treasures for which they are named—and which are formed in oysters in much the same way.

Towering columns, five feet thick and bedecked with dripstone, dominate Beth-shemesh Cave near Jerusalem. Each column represents thousands upon thousands of years of minute accretions left behind by dripping water.

A gypsum flower in Kentucky's Mammoth Cave *(left)* is composed of crystals of calcium sulfate that precipitated on porous rock. Each "petal" was forced out into the cave as new crystals formed behind it.

Crystals of aragonite deposited by slowly seeping water in a cave in France create quill-like clusters known as frostwork *(right)*. Though chemically identical to crystals of calcite, aragonite crystals differ in shape and align themselves to form more irregular structures.

Extremely rare giant quartz crystals, as much as three inches in length, abound in this Arizona cave. Some geologists believe that such crystals form underwater in flooded chambers.

A formation in Virginia's Luray Caverns owes its startling resemblance to fried eggs to a principle of optics. The large crystals of the inner circle reflect yellow light; because the smaller crystals surrounding them have more facets per square inch, they reflect a broader range of colors, and appear white.

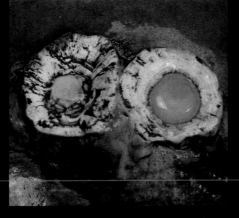

A structure that suggests the form of a butterfly, found in Sonora Caves in Texas, is the result of two kinds of water action. First, spiky crystals precipitated from seeping water; then dripping water added smooth layers of calcite over the angular framework.

Deposited in cracks in the limestone of South Dakota's Wind Cave, hard crystals of calcite remain as a weblike formation called boxwork after the more soluble limestone eroded.

Tightly coiled helictites in a cave in New Mexico seem to defy gravity as they grow upward. Water rises slowly through their minute central canals by capillary action, then deposits crystals at each tip in a wide variety of angles.

ed his other senses as well; his own rapid breathing sounded distorted.

He had prudently brought an extra tin of kerosene, but his hands trembled so as he tried to refill his lantern that he dropped the cap of the container and spilled most of the fuel. After a long struggle, he was able to relight the wick. His curiosity thoroughly quenched, White hurried back toward the entrance—and promptly cracked his head against a stalactite; blood trickled into his eyes. Reverting to a cautious pace, he picked his way along the rock-strewn corridor, found his ladder and climbed into the blessed sunlight.

The usually laconic White was now bursting with excitement, and when he got back to the ranch his story spilled out of him. Yet not even the evidence of his lacerated scalp could convince the other cowhands that he had actually made the stupendous discovery he claimed. White needed help to continue exploring—going it alone was clearly perilous—but he could persuade only a young Mexican boy to accompany him on an expedition a few days later. This time the plan was to investigate the giant cavern for three days, spending the nights beside a campfire outside the entrance.

The cowboy and his companion roamed far beyond the area that White had seen on his first trip. They followed passageways that meandered, deepened, narrowed and widened into underground rooms large enough to contain a cathedral. Other chambers were ornamented with a profusion of elaborate and beautiful formations. Pits disappeared into a blackness that defied attempts to illuminate them by lowering a lantern on a rope. Once again White was nearly done in by his crude equipment: The boy's lamp touched White's kerosene-soaked shirt and set him afire. The youngster smothered the flames with his coat, but White had been painfully burned, and their exploration was cut short.

White's report of this exploration met the same skepticism and indifference that had greeted his earlier descriptions of the cave. The skepticism may have been warranted; controversy over his claims of discovery persists to this day. By White's account, the indifference persisted for two years, until a merchant in nearby Carlsbad, Abijah Long, suddenly expressed an interest. In fact, Long may have been involved in the initial explorations of the cave, but White gives him no credit. At any rate, Long filed a mining claim in 1903, drilled a shaft into the bat chamber, and rigged a steel bucket in which bat guano was winched 150 feet to the surface to be processed into fertilizer. He and the operators who followed him during the next 20 years hauled some 100,000 tons of guano out of the cave, lowering the floor by as much as 50 feet in some places. Most of the miners strayed no farther into the cave than duty demanded, and even that could be unsettling: After a practical joker stationed a homemade ghost near one crew, the badly frightened workers quit.

White gave up his ranch job to work as a miner, but only as a means of spending more time in the cavern that had become his obsession. He continued to explore, and he even built crude trails and handrails for the visitors he hoped to someday lure to these subterranean splendors. On one excursion he climbed a steep grade and followed a passage that emerged in an awesome T-shaped chamber. White spent two days roaming this sanctum, which measured 1,800 feet by 1,100 feet, with a ceiling 255 feet high. "It was," he said, "like being turned loose out in a canyon pasture on a long, dark night with only oil torches for lights." His discovery would later be named the Big Room, and would rank for many years as the world's largest known underground chamber.

With the depletion of the major deposits in the cavern near Carlsbad after World War I, its guano production declined, and in the early 1920s the

Cowboy Jim White, who chanced upon New Mexico's Carlsbad Cavern in 1901, devoted nearly 30 years of his life to exhaustive explorations of the cave system. After Carlsbad became a national monument in 1923, White served as the cavern's first chief ranger.

Adventurous sightseers of the 1920s prepare to descend 170 feet into Carlsbad Cavern in a bucket previously used to haul up bat guano from the floor of the cave. A winch powered by a gasoline engine raised the bucket; visitors were reduced to using a rickety ladder (*right*) when the bucket was out of service.

mining operations ended. Ownership reverted to the federal government, which had no use for the cave and had no objection when Jim White stayed on, managing to lure an occasional visitor to share its pleasures.

These early tourists rode the guano bucket like an elevator down to the entrance, whereupon White would happily show them through the cave without charge. Thanks to his persistence, the number of visitors slowly increased. A breakthrough came when 13 prominent citizens of Carlsbad showed up to wait their turn for the heart-stopping ride in the low-sided bucket. With them was a photographer named Ray V. Davis, who put together a portfolio of spectacular black-and-white pictures showing flow-stone waterfalls, massive thick-trunked stalagmites, impassable thickets of bamboo-like formations, and the vast and elaborately decorated Big Room. Exhibited in Carlsbad, the photographs caused a sensation mitigated only by the still-skeptical minority who argued that pictures that fantastic had to be fakes. White, who had passed from adolescence into middle age awaiting this day, was suddenly deluged with sightseers willing to pay two dollars a head for a tour through the increasingly renowned Carlsbad Cavern.

Word of Carlsbad spread all the way to Washington and the Interior Department's General Land Office, which in 1923 dispatched mineral examiner Robert Holley to survey the cave and recommend whether it should be designated a national monument. Holley expected to spend only a short time on the assignment. Somewhat disdainfully, he told White that, although his superiors did not believe the cave was of much importance, they "thought I'd better come out and measure it so they would know if it was big enough for them to consider."

In the event, Holley spent nearly a month at Carlsbad before completing his survey and writing a report that began: "I am wholly conscious of the feebleness of my efforts to convey in words the deep conflicting emotions, the feeling of fear and awe, and a desire for an inspired understanding of the Divine Creator's work which presents to the human eye such a complex aggregate of natural wonders."

In October 1923, President Calvin Coolidge signed an order proclaiming Carlsbad a national monument, and seven years later it became a national park. The change of status brought with it, in a manner that had been beyond Jim White's scope or ken, scientific studies of the great cave.

In 1923 and 1924, extended visits by a National Geographic Society team were the first of many scientific expeditions that eventually led geologists to conclude that Carlsbad and other limestone caves in the Guadalupe Mountains had actually developed in a gigantic reef deposited more than 200 million years ago, when most of the present-day Southwest was underwater. The subterranean stresses accompanying the building of the Rockies created the system of joints that set the pattern of Carlsbad's growth. Geologists say no river or stream ever flowed through the wide corridors and imposing chambers of Carlsbad; the passageways were carved out by seeping groundwater, which gradually withdrew as the water table lowered, leaving the cave dry.

Many geologists believe that the genesis of this cave, and of others in the Guadalupe Mountains, was radically different from that of most caves. Carlsbad appears to have been created largely by sulfuric acid, formed when

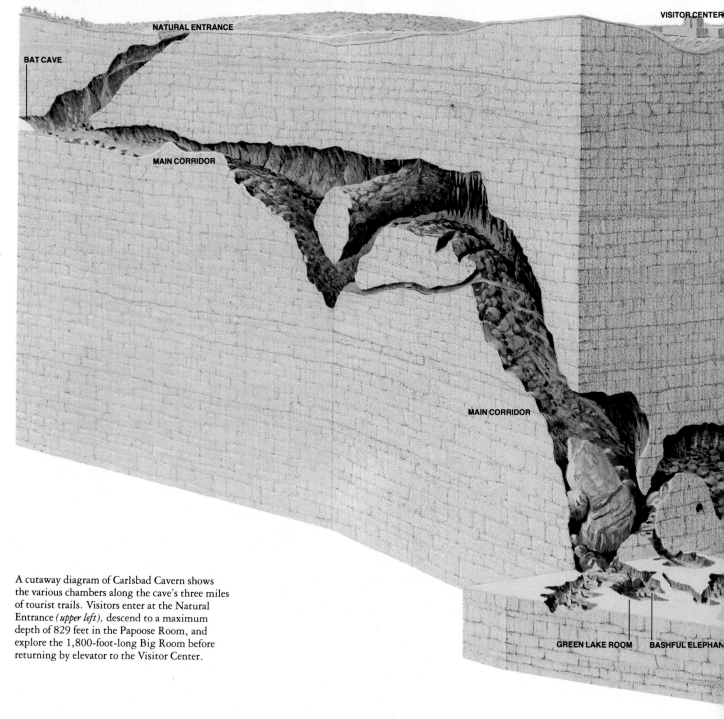

A cutaway diagram of Carlsbad Cavern shows the various chambers along the cave's three miles of tourist trails. Visitors enter at the Natural Entrance *(upper left)*, descend to a maximum depth of 829 feet in the Papoose Room, and explore the 1,800-foot-long Big Room before returning by elevator to the Visitor Center.

hydrogen-sulfide brine rose through the earth from nearby oil and gas fields. The tremendous scale of Carlsbad's chambers, with their high, vaulted ceilings, is apparently the result of the acid eating its way upward over a period of thousands of centuries. About a million years ago the first stalagmites and stalactites began to form, a drop at a time in the dry air, and they are still growing.

As for Jim White, he served the National Park Service as its chief Carlsbad ranger until 1929, when he resigned from that post in the hope of being named "chief explorer," a role he deemed more fitting to his tastes and abilities. He was turned down, ostensibly because the Park Service director felt there was no need for such a position, since "already more rooms have been discovered and explored in this cavern than can now be taken care of or shown to the visiting public."

Shunted away from the cavern he had discovered and to which he had

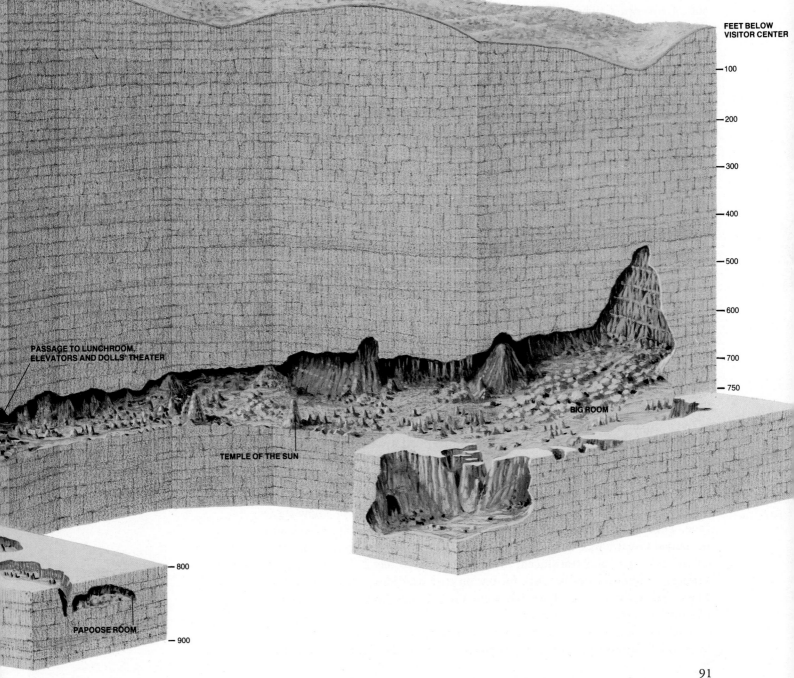

FEET BELOW VISITOR CENTER

—100

—200

—300

—400

—500

—600

—700

—750

PASSAGE TO LUNCHROOM, ELEVATORS AND DOLLS' THEATER

TEMPLE OF THE SUN

BIG ROOM

—800

—900

PAPOOSE ROOM

91

devoted so many years of his life, White died in 1946, an embittered man. Yet thanks largely to government sponsorship, his beloved cave at least was spared the indignities visited upon many other American caverns during the first several decades of the 20th Century.

During that turbulent time, the ageless caverns became the objects of greatly increased interest, not only in the United States but in Europe. The fascination took strikingly different forms.

In the United States, caves were generally exploited, either for their natural resources (as in the case of the guano mining at Carlsbad) or as commercial tourist attractions. Competition among rival cave entrepreneurs was intense and sometimes even violent. Disputes over ownership reached the highest courts in the land and, if not settled by legal decision, often resulted in bankruptcy or federal proprietorship.

In Europe, the tradition of speleology as the sporting science pioneered by Édouard-Alfred Martel was passed along to new generations—with the sporting aspects taking distinct precedence over the scientific. To the Europeans, the underground was a wonderworld of pristine beauty, of romance and of adventure, where death was to be dared if not defied. Speleological success was calculated not in terms of showplace profits but by the number of record-setting superlatives that could be applied to the discovery and exploration of caves.

Two men sum up the contrasts of the era. In Europe, the model of speleological endeavor was a scholarly Frenchman who had been attracted to caves by his boyhood reading of Jules Verne's fantastical *Journey to the Center of the Earth*. In America, the symbol of speleology was an almost illiterate Kentucky farm boy whose quest for the show-business riches of the underworld led him to the starring role in a tragic drama—a macabre cavern carnival that held the whole nation in morbid thrall.

Floyd Collins was a product of his time and place—and in a strong sense he was a victim of the bizarre hostilities that came to be known as the Kentucky Cave Wars.

The Kentucky farm folk who worked the hardscrabble land around America's original speleological frontier knew that Mammoth Cave, while indisputably the largest, was by no means the area's only formidable hole in the ground. Salts Cave, just across the narrow Houchins Valley on Flint Ridge, had been known as long as Mammoth had, and nearby Colossal was discovered in 1895. Martel himself, a 1912 visitor to Mammoth, expressed the belief that the Flint Ridge caves and Mammoth were connected and that the combined system would total 150 miles—a prophecy that proved to be conservative.

Such predictions offered glittering possibilities, but the reality was that Mammoth, still owned by the Croghan family, had a virtual monopoly on the tourist money flowing into the limestone country. Numerous and sundry would-be cave operators were determined to change that situation.

Among them, for example, was Edmund Turner, a young civil engineer who showed up in 1912 and pored over Flint Ridge in search of a new show cave. Three years later his efforts were rewarded by the discovery of a cavern he named Great Onyx Cave. Its glories included a corridor lined with giant columns of onyx and translucent alabaster and, 250 feet below the earth's surface, a huge, statue-like rock formation that was named after the Virgin Mary. Turner's dreams of wealth were short-lived, however; he died of pneumonia less than a year later.

The most successful of the Kentucky prospectors was George Morrison, a Louisville mining engineer who was convinced that Mammoth extended far

When New Mexico's Carlsbad Cavern was opened to the public in 1911, the owners made few concessions to the comfort—or even the safety—of visitors. The journey into the cave began in a singularly inelegant fashion: Tourists were lowered by guano bucket. After completing the slow, swaying descent, they were handed unwieldy kerosene lanterns to find their way about in the gloom. They quickly discovered that trails were, for the most part, inadequate: Often the tourists had to clamber over fallen rocks and creep along narrow ledges at peril to life and limb.

Improvements were gradually introduced by the National Park Service after it took over the property in 1923. Beginning in 1925, the construction of stairways put an end to the use of the guano bucket and gentled the more treacherous descents within the cave. But climbing back to the surface proved too strenuous for many tourists. By the 1950s the stairs had been replaced by smoothly paved switchback trails, which negotiate the steep slopes in a series of hairpin turns (below).

In 1927 the first electric lights were installed in the cave, and the following year an underground lunchroom began serving food in a large cavern just off the tourist trail, 1.75 miles from the entrance. Employees had to carry food and supplies in from the surface until 1931, when elevators took over the task.

All these amenities have helped turn Carlsbad into one of the nation's great tourist attractions. The cavern's breathtaking limestone formations (overleaf) are now viewed by some 800,000 tourists each year.

The ap
the flo
only o
flowst

Zigzagging trails provide an easy means of descent into Carlsbad's Natural Entrance. On summer evenings, visitors gather in the wedge-shaped amphitheater nearby to watch thousands of bats emerge from the cave.

Strategically located near the midpoint of a walking tour of the cave and about 750 feet underground, Carlsbad's modern lunchroom serves as many as 2,000 box lunches an hour.

beyond the boundaries of the Croghan land. If it did, anyone who discovered—or blasted—a new entrance outside the Mammoth tract could go into business for himself. In 1915, Morrison acquired mineral rights allowing him to probe beneath some land adjoining the Mammoth property. He made illicit surveys of the cave itself, sneaking in at night through an unused entrance on the Mammoth property. He then began drilling toward the cave on the land he had leased, explaining disingenuously that he was searching for oil. He sent men into the cave to listen for the drill and to detonate small charges of dynamite while he scanned the terrain above for telltale wisps of smoke. Mammoth officials caught him on their property once and took him to court, where he was fined $75 for trespassing.

After several years of on-and-off efforts, Morrison found what he was after: a back door to Mammoth on his own property. Morrison excavated the opening, advertised it as "The New Entrance to Mammoth Cave," built a hotel and installed a stairway to the wonders below, which turned out to include onyx draperies and waterfall formations more impressive than any in the cave's "historic" section. Even more gratifying from Morrison's standpoint was the fact that his entrance was closer to the main road—and thus more convenient to tourists—than the original one, enabling him to capture a substantial amount of business. Fending off a court challenge by the Mammoth owners, Morrison prospered for several years.

The so-called Cave Wars were precipitated during the 1920s by the bitter rivalry among Colossal, Great Crystal, Great Onyx, Mammoth Onyx, Diamond Cave, Morrison's New Entrance, the venerable Mammoth itself and others. Billboards promised "Kentucky's Most Beautiful Cave" or "The Greatest Cave of All." Agents for the various attractions roamed the major roads, sometimes flagging down motorists and leaping onto their running boards to deliver their spiels. Often they resorted to removing one another's signs or impersonating policemen to intimidate suspecting tourists (because of the visored caps they wore on such occasions, the cave pitchmen came to be called "cappers"). These uniformed impostors sorrowfully advised Mammoth-bound travelers that the tour was too long and difficult for most people, or that they had already missed the turnoff. Shills planted among the Mammoth crowds sneered at the sights and touted the superior grandeur of an attraction down the road. Fights were frequent, and a billboard-bearing truck operated by one promoter so annoyed his competitors that they burned it.

All but lost in the swarm of entrepreneurial hustlers was one man who probably knew more about caves than the rest of them put together. Floyd Collins was a lean, hollow-eyed, uneducated cave-country native whose taciturnity was infrequently relieved by a smile that displayed a gleaming gold front tooth. A member of a large family that tried with marginal success to scratch a living from 200 acres of Flint Ridge farmland, Collins from boyhood had seemed most content while burrowing beneath the earth's surface.

After years of discovering and exploring caves that were either too small or too unimpressive to be commercially profitable, he finally came by accident upon the cavern of his dreams in 1917, when he was 30. Noticing that an animal trap on the family farm had disappeared into a sinkhole, Collins probed around and found a crevice that was "breathing" air. Two weeks of hard digging uncovered a passage leading to a 65-foot-high room garlanded with hundreds of white and cream-colored gypsum flowers. Several other chambers lay beyond. Collins named it Great Crystal Cave, and had it ready for visitors a year later.

Splendid though it was, Great Crystal failed to prosper. It was, alas,

IMPORTANT NOTICE!

There is prevalent in the region surrounding Mammoth Cave a practice known as "road solicitation." Persons, often in uniform, stop visitors' cars for the purpose of soliciting business for caves other than Mammoth Cave. As a result, those wishing to reach Mammoth Cave may easily be sidetracked from their objective.

The only officially authorized places in this region giving information on Mammoth Cave are the Mammoth Cave Office in Cave City, Kentucky, located at the junction of U. S. 31-W and State Highway 70; and the entrance station of the National Park Service, United States Department of the Interior, at the entrance of Mammoth Cave National Park.

It has become necessary to warn all tourists who wish to reach Mammoth Cave to AVOID INFORMATION GIVEN THEM ON THE ROADSIDE, particularly when cars are stopped for the purpose of giving information to the occupants. The only places in which tickets to Mammoth Cave may be obtained are the Mammoth Cave Office in Cave City, Kentucky; the Frozen Niagara Hotel, located at the Frozen Niagara Entrance to Mammoth Cave; and the Mammoth Cave Hotel, located at the Historic Entrance to Mammoth Cave.

MAMMOTH CAVE OPERATING COMMITTEE

A broadside posted along the route to Mammoth Cave in the 1930s warns tourists that agents for other caves might ply them with misleading information. Mammoth Cave officials estimated that "road solicitation" by rival agents diverted up to a third of their customers.

Smiling solicitors for Great Onyx Cave, one of Mammoth's major competitors, lie in wait for gullible tourists at an information booth along the road to Mammoth Cave. The battle for customers occasionally escalated to the point of burning rival booths.

the last stop (four and a half miles beyond the Mammoth entrance) on the route known as Cave Road; tourists were enticed to rival attractions by the shills long before they reached Great Crystal Cave. Collins concluded that the only way to beat the opposition was to find another cave, with an entrance closer to the main highway than any of the others.

In January of 1925, Collins struck a bargain with three men who owned farms on a patch of high ground a few miles southeast of Mammoth Cave, where the Flint and Mammoth Cave Ridges come together. The landowners agreed to let Collins hunt for a cave on their property and to give him room and board; in exchange they would share among the three of them half of whatever profits might ensue, with Collins to get the other half.

Collins began his search in a hole he had previously spotted beneath a sandstone ledge on Beesley Doyle's farm. He spent three weeks laboriously digging a trench at the entrance and clearing debris from the narrow, twisting and downward-sloping shaft beyond. Never more than a few feet wide, the rubble-strewn crawlway extended diagonally for about 15 feet, dropped straight down for a few feet and then sloped diagonally again, doubling back under itself and narrowing to a vertical drop and a squeeze that Collins could just get through. From there it led diagonally downward again, tapering to a height of only about 10 inches before it reached an alcove barely big enough to turn around in. A few feet farther on lay a narrow, pitlike chute that dropped 10 feet to a shallow cubbyhole and a diagonal, body-sized crevice.

On Friday, January 30, a deadly drama began that was to enrapture the nation. It was the first day Collins was able to get beyond the bottom of the chute and into the crevice, thanks to a dynamite charge he had set off to clear the aperture. Carrying a kerosene lantern and a coiled rope, he wriggled through the diagonal crevice and noticed that its walls and ceiling were composed of dangerously loose dirt and protruding rocks. He emerged on a ledge overlooking a 60-foot-deep pit. Descending on his rope, he investigated the bottom of the pit until his lantern flickered, warning him that it was time to turn back. He climbed back up, leaving the rope in place for his next trip, and crawled headfirst into the body-sized crevice, squirming forward with his hips and shoulders and digging his feet into the floor and walls.

Collins was now 115 feet from the cave's entrance and 55 feet beneath the surface. Before trying to ascend from the crevice into the cubbyhole, he

him. Described in the early news accounts as weighing as much as seven tons, it turned out to be a boulder, shaped like a leg of lamb, that weighed 27 pounds.

Lee Collins sold Great Crystal Cave, the cavern that Floyd had discovered on the family farm, to a local dentist in 1927. The new owner exhibited Floyd's corpse in a glass-topped casket in the cave's main hall, a ghoulish bit of showmanship that proved to be profitable. The bitter competition that had sent Floyd Collins to his doom in Sand Cave persisted for years, spreading into such cavern-rich states as Virginia, New York, Tennessee, and especially Missouri. There, a years-long dispute over ownership and control of Onondaga Cave, 80 miles southwest of St. Louis, culminated in the placing of barbed-wire barricades along a 33-foot-wide no man's land through the middle of the beautiful cavern. The dispute was settled in 1935, but deep in the underground reaches of one of the surpassingly lovely caves in America, rusting strands of barbed wire can still be found.

While American caving developed in its unabashedly commercial way, European speleology remained a pursuit for purists. The inaccessibility and danger of many Continental caves discouraged casual visitors, yet these same characteristics posed an almost irresistible challenge to skilled explorers. One member of the elite fraternity of cavers, Norbert Casteret, spoke for them all when he declared that the ultimate sensation was to be caught "in the grip of the demon Adventure, the fascination of the unknown."

Casteret, a native of France's cave-rich Pyrenees region, grew up idolizing Martel as the man "who descended to terrifying depths to look at underground France, and who discovered the palace of the Thousand and One Nights for the greater glory of science." Like Martel, he combined a some-

Ice tinted by minerals forms an emerald drapery in Eisriesenwelt, a rare ice cave nestled high in the Austrian Alps. Frigid bedrock at the cavern's 5,400-foot altitude and air-circulation patterns that chill even warm summer air keep the limestone walls in the depths of the cave sheeted with ice year-round.

what austere and bookish manner with a robust athleticism and an appreciation for the lyrical beauties beneath the earth. As a boy, having been mesmerized by the subterranean fantasies of Jules Verne, he poked around in the numerous caverns in his home province of Haute-Garonne, learning to respect an environment where "every mistake, every act of folly is punished immediately, inevitably and often heavily." He suffered one such painful lesson during his first modest descent by rope when a lighted candle he had fastened to his hatband set fire to his hat.

Casteret's youthful explorations were interrupted by World War I. He enlisted in the French Army in 1915, at the age of 18, and was discharged in 1919. Thereafter he undertook the study of law, a calling that he soon discovered was "not at all adapted to my temperament, which was sportive, venturesome and hardened by war's school of iron and fire." Casteret decided to make caving his livelihood. As he admitted later, the decision "seemed madly ill-considered," but it would turn out to be of enormous consequence for 20th Century speleology.

Throughout Casteret's long career, his motivating force was the exploration, not the exploitation, of the wonders that lay underground. In spite of professional attainments that exceeded even his fondest hopes, he would always remain a caving purist. For example, many years after sturdier and more stable craft had been developed, Casteret, "as a sort of tribute to my illustrious predecessor, Martel," insisted on negotiating underworld streams in the same kind of collapsible skiff that Martel had used.

Before he would be able to move ahead in his fledgling profession, Casteret clearly needed to achieve a degree of speleological distinction—and his opportunity arrived in 1922 when, at 25, he was making a systematic study of Pyrenees caves.

A motionless cascade of ice blankets a slope in Eisriesenwelt, or "world of the ice giants." The icy ramparts, which presented a formidable barrier to early explorers, now attract thousands of tourists to the cavern.

Chapter 4

MASTERS OF MAKING DO

Deep within the underworld prevails a silence of the tomb. Beneath the shroud of eternal darkness, no green plants—the basis of the food chain for most creatures—can grow. The environment could hardly seem less hospitable to life. And yet, astonishingly, animals can thrive in the caverns of the earth. Unseen, unheard, they creep and crawl, slither and scuttle, swim and fly, endlessly questing for the means of survival. In their marvelous physical adaptations, in the senses they use to navigate the perpetual night of a rockbound realm, in their extraordinary behavior and reproductive patterns, they offer a microcosmic study of evolution.

Unlocking the secrets of cave fauna is the objective of a fledgling science called biospeleology. Armed with instruments ranging from ordinary dip nets to highly sophisticated electronic equipment, biospeleologists venture into the planet's bowels in search of the animals that eke out a precarious existence there. The esoteric pursuit has led to the discovery of many strange new species—and of variations within species that are just as strange.

As a first step toward imposing order on their science, biospeleologists place cave creatures in three categories. Troglophiles, or cave lovers, is the name given to species whose members are capable of going through their entire life cycles either within caverns or on the earth's surface; several varieties of spiders and crickets are in this category. A second group is the trogloxenes, or cave visitors; these creatures—bats and cave-nesting birds, among them—customarily live in caverns, taking advantage of such comforts as constant temperatures and the absence of predators, but from time to time they must leave to forage for food. And the hard-core, full-time cave inhabitants that are never found outside their underground habitats are called troglobites, or cave dwellers.

The detection, classification and study of these otherworldly creatures is hardly an enterprise for the fainthearted, as an expedition into a cave near Springfield, Missouri, in 1938 attests. Charles Mohr, who would later become president of the U.S. National Speleological Society, and a colleague, Kenneth Dearolf, were determined to find an Ozark blindfish, a species that had been the subject of intense scientific curiosity since its discovery in the late 19th Century by a Missouri caver named Ruth Hoppin. With a rough-and-ready kind of experiment, Hoppin had determined that the small (five inches at the longest), white, catfish-like creatures were not only blind, but deaf as well. "I tested their hearing by hallooing, clapping my hands and striking my tin bucket when they were in easy reach

A stilt-legged harvestman, popularly known as a daddy longlegs, straddles stalactites in Spruce Run Mountain Cave in Virginia. During the winter, this spider-like creature lives within the cave; in the summer, it forages for food in the woods outside.

and near the surface," she wrote, but "in no case did they change their course or notice the sound."

Mohr and Dearolf began their pursuit of the blindfish by dropping into the Missouri cave through a brush-choked entrance hole and working their way down a 30-foot chimney to the top of a 25-foot-high waterfall. While Dearolf remained above, Mohr lowered a rope and descended hand over hand at the edge of the cascade. The pounding water forced him to let go of the rope partway down, but he landed intact. His scientific curiosity quickly overcame the shock of his fall, and he began searching a muddy pool in a chamber behind the waterfall. After gathering up two crayfish, he spotted exactly what he was after—a two-inch-long fish with puffs of fatty tissue where its eyes should have been. He tried to scoop the prize into his cupped hands, but the fish, although obviously blind, easily avoided his lunge.

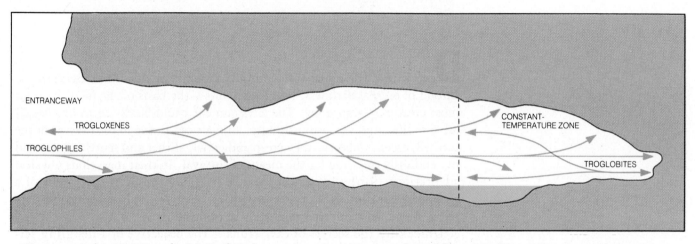

Preoccupied with his task, Mohr failed to realize that the roar of the falls had become louder and the water in the pool was rising. A sudden cloudburst on the surface, unheard by the men belowground, was pouring torrents of water into the cave. Dearolf gestured upward urgently: time to get out. Hurrying to the rope, Mohr was dismayed to discover that its lowest loop was shoulder high; try as he might, he could not pull himself up against the waterfall far enough to reach the loop with his foot. He shouted over the clamor of the water to Dearolf, telling him to get a longer rope from a nearby farm. Then he retreated to the chamber behind the waterfall, where he nervously eyed the rising water and swung his arms to try to keep warm.

After what seemed like an hour, Dearolf returned and dropped a longer rope. Mohr began to pull himself up through the avalanche of water. The force of the rain-swollen cataract jammed his head lamp down over his ears. About halfway up, exhaustion and the pressure of the roaring water compelled him to stop. For a few moments he dangled helplessly; then he began to swing back and forth until he fetched up on a ledge. There, away from the waterfall's full force, he gathered his strength for a moment, and launched a desperate effort. With Dearolf hauling from above, Mohr climbed frantically, reaching safety at the top with his last strength. Later, as the two men crossed a meadow on their way back to their car, they noticed drowned chickens floating by in a swollen river that a few hours before had been a trickling creek.

Three weeks after this harrowing experience, Mohr and Dearolf returned to the cave. They placed a beam several feet from the falls, suspended ropes from it and made an uneventful descent. A blindfish was swimming imperturbably in the same pool where Mohr had seen one before, and this time he got it. The captive proved to be identical to the fish studied

The habitats of the three types of cave animals are diagrammed at left. Trogloxenes, or cave visitors, are temporary residents and get their food mainly from the outside world. Some troglophiles, or cave lovers, complete their life cycle inside, but others of the same species live outside the cave. Troglobites, or cave dwellers, can survive only in a deep cave's climatically stable zone.

by Ruth Hoppin 50 years earlier in a cave more than 50 miles away.

Scientists have learned a great deal about adaptation from such encounters with troglobitic blindfish. Even more revealing are discoveries of troglobitic and troglophilic fish swimming together. A notable event of this type took place 20 years after the Mohr-Dearolf expedition. The place was Kentucky's Mammoth Cave; the protagonists were the surface-dwelling troglophile *Chologaster agassizi,* or springfish, and the blind troglobite *Typhlichthys subterraneus,* or southern cavefish.

During floods, the cavern's River Styx draws water from the Green River outside. When the water is high, springfish invade the cave in considerable numbers, and since the flood also brings with it food supplies in the form of vegetation and other floating edibles, the troglophilic fish prospers along with its cloistered counterpart. However, during those years when the Green River and, consequently, the Styx fail to flood, the cavefish, accustomed to making do on very little, becomes king of the stream while the springfish starves and disappears until the Styx is replenished by new floods.

In 1958 a team led by biospeleologist Thomas L. Poulson entered Mammoth Cave after the highest spring floods ever recorded there: The water rose 60 feet above the low-level mark. Almost immediately they found a springfish and a cavefish in the same pool. The differences between the two were obvious even to an untrained eye: The springfish was brown, was built like a snub-nosed bullet and possessed small but otherwise normal eyes and short fins. The cavefish, on the other hand, was almost colorless, with a large head, mere bulges where its eyes should have been, a more sinuous body and long, almost diaphanous flowing fins.

Neither fish seemed inclined to exert itself unless necessary. But when seeking food or threatened, the springfish was a more vigorous swimmer, darting busily about with short, sharp thrusts of its stumpy fins. In contrast, the cavefish took a noticeably straighter course and coasted along with smooth fin strokes. Each fin flick of the cavefish carried it about three times the distance covered by the springfish's equivalent effort.

Both fish resisted capture for a time. The springfish was fast; whenever Poulson brought his net close to the fish, it darted wildly away. The languid movements of the cavefish, on the other hand, made it easy to approach, but—despite its blindness—it clearly sensed the approaching net and dived to avoid it.

When Poulson examined the two kinds of fish in the laboratory, he found a marked difference in their sensory capabilities. Along the head and body of both types of fish run lateral ridges, which are actually rows of sensory organs known as neuromasts; from each neuromast extends a tiny, jelly-like rod, called a cupula, so sensitive that it responds even to the water movement caused by the approach of a microscopic water flea. To its great advantage in the darkness, the cavefish possesses about four times as many neuromasts as the springfish and its cupulae are twice as sensitive. With such equipment and a smoother swimming style that causes less water disturbance, the pallid, troglobitic cavefish is significantly better suited than the springfish to sense the minute underwater movements of the water fleas and tiny crustaceans that provide sustenance.

In a cave, all food—except, of course, a meal in the form of another cave animal—must somehow be imported from the outside. For that reason the troglobites, imprisoned in caves by their own nature, must rely almost exclusively during most of the year on the more worldly troglophiles and trogloxenes, which, directly or indirectly, deliberately or by inadvertence, transport provender into the underworld.

121

physical, sensory or behavioral characteristics particularly suiting them to life underground. The blind salamanders that crawl lazily through the waters of widely separated European and American caves admirably support this idea: Whether on the surface or below, salamanders are endowed with low metabolisms; they require moisture to breathe, making caves, with their high humidity, an especially friendly habitat; and even surface salamanders prefer dark places beneath rocks or in mud and slime. But it is in the differing evolutionary stages of salamanders found in caves far removed from one another that the reclusive animals are of most interest to biospeleologists.

The first scientific mention of a cave-dwelling salamander appeared in a book published in 1689 by the Slovenian Baron Johann Valvasor. A caving enthusiast when he was not bearing the burdens of title, the baron spotted the creature in a stream in a Yugoslavian cave. Pinkish white, eyeless, perhaps a foot long, supported by an odd set of front feet with three toes and by hind feet bearing two toes each, this unlikely beast looked as though it could be a new and pint-sized species of dragon. Local villagers had long believed that a dragon living in a cave at the source of the River Bella caused periodic floods by opening sluice gates when its living quarters were threatened by rising water.

The truth, as it gradually emerged over the next two centuries, was hardly less fantastic. The miniature dragon was actually a salamander of the type that came to be known as *Proteus,* the name of a sea-god capable of changing its form. Born with rudimentary but sightless eyes, *Proteus* loses them as it matures, until not even the sockets are recognizable. It changes from a dark gray color at birth to a pigmentless white, tinged slightly by its blood vessels. It possesses both functional lungs and a set of feathery red gills on each side of its neck. *Proteus* appears to be sensitive to light in spite of its blindness, although precisely how remains one of its many secrets.

So strange an animal was worthy of even royal attention: Archduke Jean of Austria kept a specimen in a grotto at his country house for eight years. Formal scientific recognition and a dignified name *(Proteus anguinus)*

White fungus flourishes on a thick, decaying layer of nitrogen-rich bat guano that blankets the floor of a cave in Missouri's Ozark Mountains. Fungi are among the few forms of living vegetation that contribute to the sparse food chain in caves.

A tiny millepede, eyeless and devoid of pigment after eons of adaptation to the perpetual darkness of a cave, somehow senses a beam of light and curls up in a frantic effort to avoid it.

stripped *Proteus* of its reputation for having dragon powers, but made it even more interesting to scholars, many of whom obtained one for their laboratories. These lab specimens astonished scientists by demonstrating their ability to survive for extended periods of time without apparent nourishment: One collector avowed that his *Proteus* succumbed to starvation only after 14 foodless years. Modern students of *Proteus* are skeptical of such claims, but suggest that a mature specimen in a cool environment could survive as long as three years without any food whatever.

Later investigators added more pieces to the still-incomplete portrait of this bizarre creature. During the mating season in May—by no coincidence a month with a high flood rate—the male *Proteus* establishes a territory and challenges any trespassing male, flailing away with its tail and biting the intruder if it does not leave; how *Proteus* senses the alien presence is unclear. It was once believed that in cool temperatures *Proteus* bore live young like a mammal, but that it laid eggs like a reptile if the water was warmer than 59° F. In 1958, however, biologists were finally able to prove that the females always lay eggs, usually under large underwater rocks, regardless of the temperature. Unlike its relatives outside caves, *Proteus* does not metamorphose from a larva to an adult form, but retains a larva-like form throughout its long life of up to 25 years. Its senses of touch and smell are apparently keen. No one has yet fully explained exactly how or why its eyes deteriorate beneath the skin as it matures; eye development seems to stop just before *Proteus* hatches and to regress afterward.

The first clue that the European *Proteus* had an American cousin came in an 1885 *Scientific American* article that included "white, blind lizards" among the fauna found in Missouri's Marble Cave. Six years later, a single specimen of what would come to be known as the Ozark blind salamander was plucked from another Missouri cave and dispatched to the Smithsonian Institution. Others were subsequently found, and scientists learned that the Ozark blind salamander differs from *Proteus* in several striking particulars. The Ozark salamander, for example, makes occasional forays to the outside world, apparently because it lacks the ability of *Proteus* to survive for long periods on little or no food. Even more significantly, the Ozark species metamorphoses from a fairly dark, relatively stocky creature with small but functioning eyes to a slender, pale and sightless adult: Its eyelids grow together as it matures.

As it happens, the process of metamorphosis requires a lavish expenditure of energy; it is therefore a considerable disadvantage in the world of caves, where the food needed to generate energy is in such short supply. In that major respect, *Proteus* is better adapted to cavern conditions than the Ozark blind salamander. If, as scientists assume, the length of time an animal species has spent in the caves can be gauged by the number and degree of cave-adapted features, it seems likely that the ancestors of *Proteus* were veterans of the underground when the forefathers of the Ozark blind salamander were still hiding under rocks on the surface.

Another American member of the family was discovered in 1895, when a crew of well drillers near San Marcos, Texas, saw several skinny white salamanders, of a sort previously unknown, rise to the surface from a stream they had tapped 188 feet underground. Only a few more of the creatures were subsequently seen—until 1938, when the ubiquitous team of Mohr and Dearolf set out to look for them in Ezell's Cave, near San Marcos, where several had reportedly been sighted.

The two men lowered themselves 40 feet into the entrance chamber by rope, then inched their way through a 100-foot-long network of narrow

THE HARDY DENIZENS OF DARKNESS

"Natural selection is daily and hourly scrutinizing, throughout the world, every variation, even the slightest; rejecting that which is bad, preserving and adding up all that is good; silently and insensibly working, whenever and wherever opportunity offers, at the improvement of each organic being in relation to conditions of life."

This observation by Charles Darwin, published in 1859 in *The Origin of Species,* applies nowhere more accurately than to the innermost reaches of caves and to the troglobites, the subterranean fauna that inhabit them. No creatures on earth are better showpieces of his theories or offer more dramatic demonstration of the basic Darwinian tenet that only the fittest survive.

All troglobitic creatures—including certain species of fish, insects, crustaceans and amphibians—evolved from surface-dwelling ancestors whose environments resembled, in some respects of humidity or darkness, that of caves. Long ago, some of these creatures began to colonize caves. After ages of genetic isolation and adaptive change, their progeny became permanent residents.

The successful creatures in this realm tended to be smaller, with slower metabolisms and longer life spans. Their eyes degenerated, since the perpetual darkness yielded no advantage to those who had sight. With no need for protection from the sun or camouflage from enemies, the cave creatures lost their pigmentation. They developed elongated limbs and fins capable of more efficient movement, and highly sensitive sensory organs to detect the presence of predator or prey.

The ghostly Texas blind salamander *(right)* and the other troglobites shown on the following pages epitomize the legions of creatures that—because of the process Darwin so eloquently described—can never return to the surface of the earth, but are wonderfully equipped to survive below.

Breathing through feathery pink gills, a four-inch-long Texas blind salamander turns its spoonlike snout in search of prey, which it finds with highly developed sensory organs along its sides and tapering head.

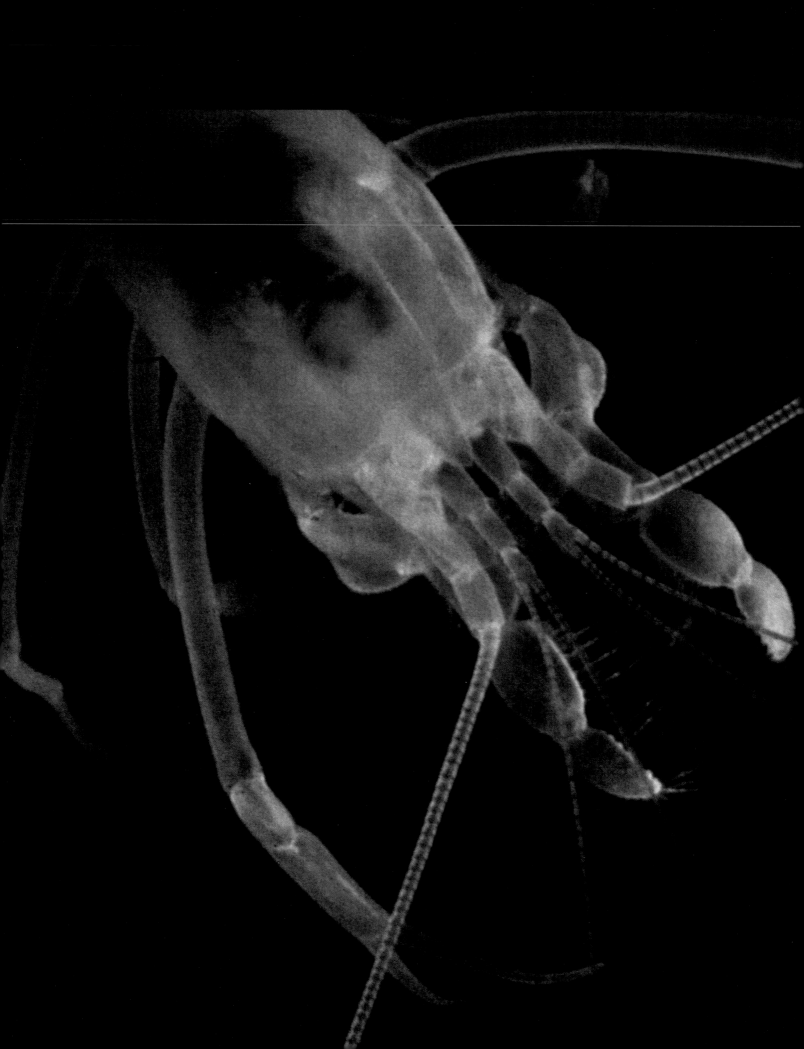

A cave cricket, clinging to a cavern ceiling with stiltlike legs, displays antennae that are twice as long as those of surface crickets. Unlike its chirruping relatives, the cave cricket is mute.

A close-up view of an inch-and-a-half-long, translucent cave crayfish shows the unique hairy feeding appendages that jut forward from its head. These oversized limbs are used to brush food, collected on the long legs and antennae, into the crayfish's mouth.

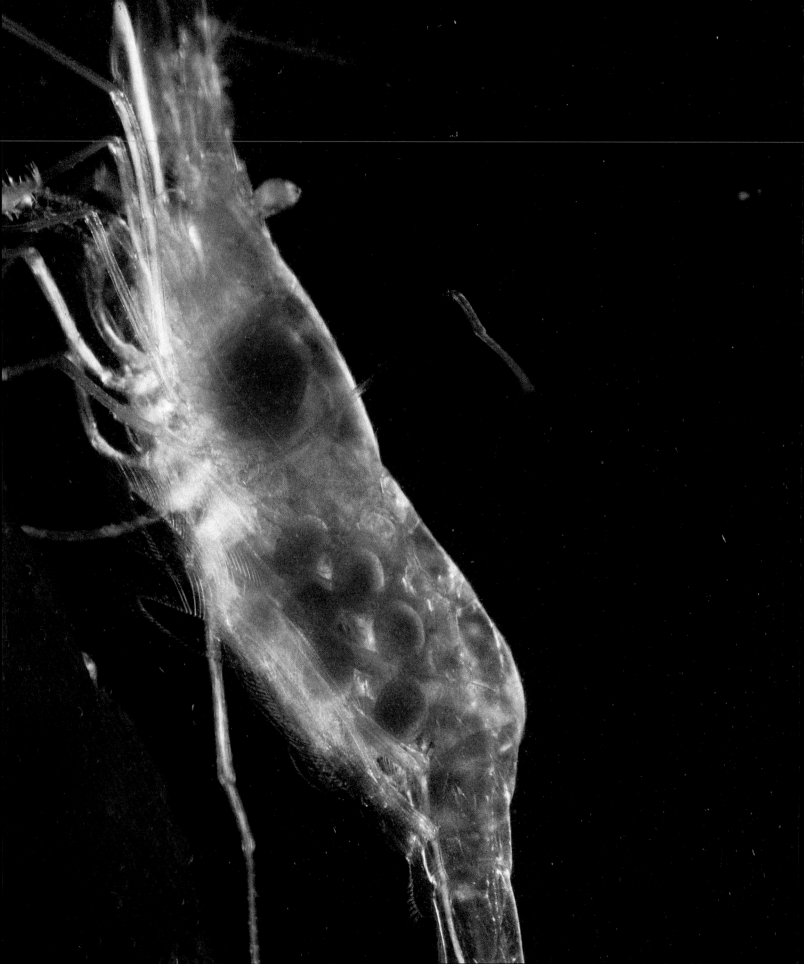

This rare blind shrimp is found only
in Squirrel Chimney, a flooded sinkhole near
Gainesville, Florida. The cave-dwelling
shrimp produces fewer eggs than other shrimp
varieties, but because the yolks are larger,
the embryonic blind shrimp are well nourished
and well developed when they hatch.

Even inside her web, a troglobitic spider keeps
her eggs on her back. Most spiders living
aboveground deposit their eggs on a convenient
surface, but closer guarding of the eggs is
necessary in the subterranean world because cave
predators are so thorough in their quest for food.

141

Marcel Loubens *(center)* enjoys an animated chat with a Pyrenean shepherd *(left)* during an aboveground respite in the 1951 explorations of the Cave of Pierre St.-Martin. Listening in are Georges Lépineux, the cave's discoverer *(right),* and expedition member Haroun Tazieff.

into clubs and federations. Caving societies began to blossom throughout Europe, and in 1941 the National Speleological Society was chartered in the United States. Within a decade, well-organized expeditions had replaced the solitary adventurings of the previous generations.

These expeditions were both systematic and purposeful. Using compasses to record the direction of movement, clinometers to measure angles of descent and ascent, and altimeters to determine depth, the teams of explorers and scientists meticulously charted every passage they found. Expeditions returned to the same cave year after year to probe promising leads.

The improved organization, methodology and technology of caving amounted, by 1950, to a thorough transformation of the art of underground exploration. It was a transformation that would make dramatic contributions to scientists' understanding of how caves form, how they change and how they have come to be the repositories of some of nature's most dazzling geologic treasures. At the same time, the known boundaries of the world's underground environments would be greatly expanded, most notably in such sprawling cave networks as Hölloch Cave in Switzerland and Kentucky's Flint Ridge system. Inevitably, these successes further whetted cavers' appetite for the singular combination of danger, beauty and conquest offered by the underworld frontiers.

The expeditionary approach to caving was still in its infancy in 1951 when 12 European cavers gathered in a remote section of the French Pyrenees for a team venture into an intriguing sinkhole. The previous year, one of their number, Georges Lépineux, had been searching for caves in the limestone mountain range when he spotted a crow emerging in full flight from what appeared to be a blank rock face. Looking closer, Lépineux found a small, concealed opening to a seemingly bottomless pit. He contacted his friend Max Cosyns—a Belgian scientist and a veteran caver—and Cosyns organized the assault. Among the men Cosyns enlisted were French volcanologist Haroun Tazieff and Marcel Loubens, a 28-year-old protégé of the eminent speleologist Norbert Casteret. For the initial descent into the shaft, Cosyns planned to use a pedal-powered winch and more than 1,300 feet of 1/5-inch steel cable.

Lépineux's sinkhole, called the Cave of Pierre St.-Martin after an old border post nearby, opened onto a rugged, wind-swept ridge at an altitude of 5,660 feet near the Spanish border. Here, long ago, the limestone rock had been compressed and folded by titanic mountain-building forces into a landscape riddled with deep vertical faults. Over the years, the substantial rainfall of the region had carved out spectacular chimneys leading to large underground chambers. Through these channels still coursed millions of gallons of water daily, bursting in powerful cascades from the cliffsides of nearby gorges.

Cosyns' group set up camp on the barren hillside and anchored the winch near the entrance to the shaft. Lépineux, as the discoverer, was the first to descend. Wearing a tight, rib-bruising parachute harness attached to the cable by a single metal clamp, Lépineux backed slowly over the edge. The crewman on the winch pedaled steadily, paying out the wire. Soon, Lépineux passed beyond earshot, and thereafter relied on a telephone line for communications with the surface.

Walking slowly backward down the wall of the shaft, Lépineux found a sloping outcrop ledge at a depth of 260 feet. Some 440 feet farther on, he descended through an icy waterfall that spilled out of a fissure in the wall. Drenched and bone-cold, he continued the descent, loosing showers of rocks as he negotiated a series of outcrops. After about an hour and a half, he found himself twisting slowly in open space as he was lowered through the roof of a large chamber whose floor appeared to be about 150 feet beneath him.

His chest aching from the brutal pressure of the harness, Lépineux finally touched down more than 1,100 feet below the surface—a new world's record for vertical descents. He scrambled down a rocky slope and explored the chamber for several hours, until fatigue drove him back to the winch line for the tedious return journey to the surface.

The next day, Tazieff and Loubens went down and started to explore in earnest. Beyond the area Lépineux had seen, they came to a small hole in the floor of the chamber; through it, they could see a terrace and another pit. Loubens explored the lead briefly and announced that a great chasm had been found, but this route down was too dangerous. Hours later, they found a safer descent. Loubens tied himself to a rope and climbed down into the passage, using the wire ladder they had brought along. Soon his voice rose from the gloom: "It's huge, positively huge. I can't see the walls. I'm going down a little farther."

Tired, alone in a darkness that seemed to swallow his feeble light, Tazieff settled back to wait, knowing that they must report back to the surface by phone in less than three hours. After a very uncomfortable hour, Tazieff shouted into the hole. There was no response. Another anxious, silent hour crept by. Finally, he heard and joyfully responded to a faint cry from Loubens. Both of Loubens' lights had gone out, but with frequent shouts and a brilliant magnesium flare, Tazieff managed to guide the exhausted but exhilarated explorer back. "Lost myself," Loubens explained nonchalantly, adding, "This is a *cave*." A little while later, the tension and fatigue caught up with him. While rolling up the ladder after a climb, Loubens suddenly broke down and wept. "I was really frightened," he confessed.

Although he had flirted with disaster by going on alone, Loubens had found not only a large gallery, but also a caver's most sought-after prize—a river that undoubtedly led to more cave. Bearing this exciting news, the two men hurried back to the shaft, and soon afterward they were on their way up and out. Encouraging though their tidings were to the other team members, Cosyns declared that the winch was about to give out and called a halt to the expedition for that year.

The following summer, most of the original party returned, including Loubens and Tazieff, and this time they were joined by the French speleologist Norbert Casteret. Plans called for the men to be lowered into the pit quickly with an electric winch, but the machine was beset by gremlins, and delays were frequent. Loubens and Tazieff spent the whole of the first day getting to the bottom of the shaft and setting up camp. The following day, they surveyed the chamber that Loubens had discovered and saw that the river disappeared through an impassable opening in the rock. After a 10-hour search, they finally found a shaft that revealed a second large chamber below. Too tired to go on, they returned to camp, where they were joined by two other team members. With these reinforcements, they found a passage around the river siphon on the third day, but again exhaustion caught up with them before they could explore farther. That evening, Loubens announced that he would return to the surface to give someone else a chance. He phoned Casteret. "I've had my share of fun," Loubens told him. "I'm all in."

In the morning, the men helped Loubens into his harness and watched as he trundled up the pile of boulders they used as a lift-off platform. A few minutes later the cable stretched taut and he began to rise, spinning slowly. When he was about 35 feet up, he tried to ignite a flare so that Tazieff could take a photograph, but the matches blinked out one by one in the strong draft.

Still hoping for a picture, Tazieff was squinting up through the camera's viewfinder when he saw the beam of Loubens' head lamp suddenly plummet toward him. At the same time he heard a cry. An instant later, Loubens struck the boulder Tazieff was standing on and rolled down the rock pile, bouncing from stone to stone. Another team member finally halted his fall 100 feet farther down the slope.

Loubens was unconscious, breathing with rapid gasps. With infinite care on the perilous rubble slope, the men eased him onto a sheet of canvas and carried him back to the campsite, the only place where he could lie flat. While Tazieff removed Loubens' helmet and examined his skull for signs of fracture, one of the men scrutinized the dangling cable. At the end of the line he discovered the cause of the fall: The metal clamp that attached the harness to the cable had worn through.

They immediately phoned the expedition doctor, André Mairey, on the surface. He said that he would come down as soon as the cable was hauled up and the shackle repaired. In the meantime, the others could only wait and watch their friend's torturous efforts to breathe. Occasionally, they wiped his face with a damp cloth. Hours passed with no sign of Mairey, and then the phone inexplicably went dead. The surface crew labored to repair both the clamp and the phone, but progress was agonizingly slow. That night, a thunderstorm struck the topside camp with powerful winds, further delaying the doctor's descent. When Mairey finally reached the bottom with a stretcher, Loubens had been fighting for breath for nearly 24 hours. The doctor leaned over and examined him carefully. Loubens had a broken back and a fractured skull: His chances of survival were infinitesimal. But as long as he remained alive, the men were determined to try to get him out.

While Mairey and the men below strapped Loubens into a harness and bound him tightly to the stretcher, the topside crew linked together several 65-foot-long ladders and dropped them into the pit. Casteret and four others climbed down the ladders and took up positions at different depths on narrow ledges, anchoring themselves with pitons. The lowest man stationed himself 790 feet below the surface. Each was prepared to risk his life to shepherd the stretcher upward past the outcrops in the shaft. Far below,

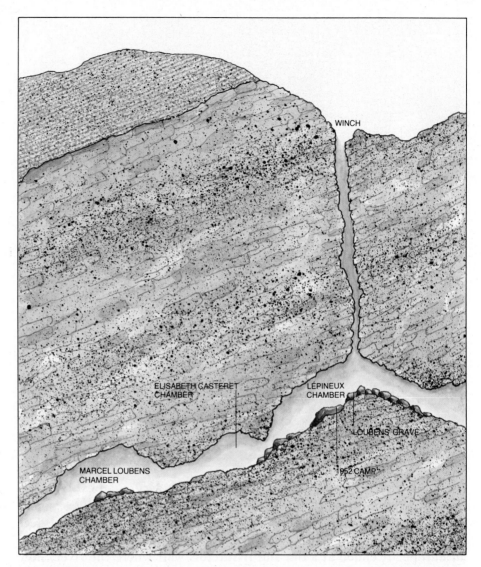

In the Cave of Pierre St.-Martin, a 1,100-foot shaft opens into a series of spacious underground chambers. After Marcel Loubens' fatal fall at the bottom of the shaft, his companions entombed him near their subterranean camp in a crypt of loose boulders.

rocks dislodged by the rescue team peppered the camp as Mairey gave the injured man a blood transfusion.

Mairey was still awaiting the go-ahead signal when he heard Loubens groan. Seconds later, almost 36 hours after his fall, Loubens' labored breathing stopped. The surface team received the solemn announcement by phone: "Marcel Loubens is dead." The word was passed to the men in the shaft, and one by one they climbed back up. The risks they were prepared to take to save an injured comrade were deemed too extreme simply to remove his body.

The next day, the men in the cave wrapped their friend's body in canvas and laid it in a hollow between two boulders, covering it with gravel and stones. Tazieff fashioned a small cross out of sheet metal. Then, using a hammer and chisel, the men carved an inscription into the rock near the grave: "Here Marcel Loubens passed the last days of his gallant life."

Although heartsick, Tazieff and Mairey decided to stay below long enough for a last look at the passage Loubens had found at the end of the second chamber. A series of twisting corridors brought them again to the elusive river. Making their way through a small room, they emerged into the main hall, a huge, smooth-floored vault decorated with soda-straw stalactites and large mushroom-shaped stalagmites. Tazieff lighted a flare and marveled at the chamber's grandeur—it was the largest room yet. At its low end, the room pinched down to a flooded tunnel. Here they decided to turn back, but first they named the spectacular cavern Loubens Chamber.

Cavers and bystanders cluster on the mountainside above the entrance to the Pierre St.-Martin system in 1952 to celebrate a mass in memory of Marcel Loubens. His death a week earlier at the bottom of the shaft curtailed the preliminary exploration of the cavern.

Casteret, at the age of 56, spearheaded the underground explorations when they resumed in 1953. Teams led by the veteran speleologist found four more large chambers. The last of the four measured some 660 feet by 390 feet, with a ceiling 330 feet high—the largest cave chamber in Europe. It ended at a flat and fissureless wall where the river, now merely a trickle, seemed to vanish into the floor. According to the cavers' altimeter, the gallery was deeper than any yet discovered—2,389 feet below the surface.

But for the Pierre St.-Martin explorers there remained the melancholy task of bringing Loubens' body back to the surface. The following year, they placed his remains in a coffin-like aluminum container and, wrestling it past the outcrops in the shaft, managed to haul it up with a motorized winch.

The sustained teamwork at Pierre St.-Martin, though shadowed by the death of Loubens, remains a triumph in the annals of caving. The expeditions demonstrated the merits of organization, pioneered a promising vertical technique and explored deeper into the earth than anyone had before. The same team approach to caving was used by other groups that came to Pierre St.-Martin in the 1960s and 1970s and discovered two more routes beyond Casteret's last probe.

While Cosyns and Casteret struggled with their winches at Pierre St.-Martin, American cavers in the Appalachian Mountains were experimenting with different methods for vertical caving. Instead of moving the line itself up and down from above, they used techniques that allowed the individual climber to control his own movements on a stationary vertical rope. At first, the cavers tried the movable prusik knot for one-man ascents, and for descending used the rappel, a method of walking backward down a cliff face, controlling the rate of descent by paying out rope gradually through load-reducing knots and fittings.

More imagination was required for descending into pits when contact with the walls could not be maintained. The solution was the rappel rack, a device that permits the caver to control a descent by weaving the rope around several horizontal braking bars to increase friction. The metal bars are movable: To descend faster, fewer bars are used or the spacing between them is increased; to go slower, the bars are pushed together or more bars are used to increase the purchase on the rope. American tinkerers also developed mechanical versions of the prusik knot. Like the knot, these mechanical ascenders clamp the rope tightly when bearing weight, but they slide upward when the weight is removed.

Vertical techniques were improved and refined in North America for almost a decade before an opportunity arose to give them a truly dramatic field test. In the mid-1960s, a group of American cavers took their gear to Mexico's Sierra Madre Oriental, site of a 1,092-foot hole called the Sotano de las Golondrinas. Here, they would have to move down and up a rope near the middle of the shaft, far from the walls. Using rappel racks, they made the spectacular "free-fall" descent in only 30 minutes, less than one third the time taken by the men on the winch in Pierre St.-Martin. Although the climb back up with the mechanical ascenders took about two hours, the cavers were able to rest en route; they used a third ascender attached to a chest harness to keep themselves balanced and upright while they caught their breath.

The technical innovations had made possible an entirely new kind of caving experience. As one Golondrinas veteran remembered: "I was suspended in a giant dome with thousands of birds circling in small groups near the vague backcloth of the far walls. Moving slowly down the rope, I had the feeling that I was descending into an illusion and would soon become part of it as the distances became unrelatable and entirely unreal."

Not all the changes taking place in speleology were dramatic: One of the most important trends was a new appreciation of carefully designed research plans and systematic surveys. The first great proponent of this approach was Swiss geologist Alfred Bögli, who supervised the investigation of Switzerland's Hölloch Cave in the early 1950s.

Hölloch Cave wanders for miles beneath the Swiss Alps, and for many years its only known entrance was near the bottom of a 2,800-foot series of passages and jagged chambers. Although Hölloch was discovered in 1875, only four miles of it had been traced when Bögli began surveying the area in 1945. Six years later, he became scientific director of a group exploring the cave. By then, a large iron gate had been built across the entrance to keep out interlopers, a move Bögli had reason to regret the very next year. In August of 1952, a flood marooned him and three teen-age assistants in the subterranean depths. "We were in a room with no flat place to lie down on, and the temperature was 41° F.," he recalled. For nine cold and tedious days, they rationed their food—"600 calories a day, mainly soup and bread"—and waited for the water to recede. On the 10th day, Bögli felt a draft, a sign that water no longer completely filled the passageway. He and the three boys navigated a siphon while holding on to a rope and made their way back to the entrance, where they found that their would-be rescuers had gone home and left the gate locked. They had to force the gate open to get out.

Undaunted by the incident, Bögli returned each year to Hölloch during the one fortnight in December when the cave was dry enough for exploration. By 1955, he and his crews had mapped some 34 miles of the cave.

After lying for two years in the chill depths of Pierre St.-Martin, Marcel Loubens' body is hauled to the surface in 1954. Because of frequent snags on the walls, winching the aluminum coffin up the shaft took 20 hours.

Emboldened by their success, Austin and Lehrberger joined several other experienced cavers to form the Cave Research Foundation, professing scientific aims in an attempt to legitimize their efforts. The Park Service soon decided that an organized group of disciplined, experienced cavers could be a great asset to the park, and in 1959 the foundation members were granted full access for exploration. With this franchise, they revived the expedition approach, and it paid off almost immediately.

Lehrberger and two other explorers were dispatched in 1960 into Colossal Cave. They quickly pushed beyond the known trails and followed a stream to a passage so low that they had to crawl on their bellies through the cold water. However, a little farther on, they climbed a ledge and found themselves in Salts Cave—thereby establishing another new Flint Ridge connection.

According to a survey map, only 160 feet separated the Colossal-Salts paired system and the Unknown-Crystal pairing. In the summer of 1961, two other Cave Research Foundation explorers felt a draft coming through a rock pile in Unknown Cave and clawed at the rubble until they made an opening large enough to crawl into. They struggled through and emerged 30 minutes later in Salts Cave, thus demonstrating that all four Flint Ridge caves were in fact one 30-mile-long system. The explorers once again cast longing looks across the broad Houchins Valley toward Mammoth.

During the next three years, teams of cavers systematically explored the avenues of the Flint Ridge caves that led toward the valley. Finally, in 1964, they discovered a long crawlway that projected under the north end of the valley. Several exploratory missions extended this trail to an area only 2,000 feet from Mammoth Cave.

The following year, another team pushed on toward Mammoth until they were stopped by a massive breakdown pile beneath Mammoth Cave Ridge just 800 feet short of the cave. Team members tried repeatedly to clear an opening through the breakdown, but every time they removed one rock, another tumbled into its place. Finally, they gave up, and the hope of using this route to unite the two gigantic cave systems faded and all but flickered out. Only a special kind of speleological stubbornness kept it alive.

The Flint Ridge scramblers continued to probe every opening they could find. None paid off. It was scant consolation that by 1969 their dogged surveys had earned the Flint Ridge system the title of the world's longest cave, surpassing both Mammoth and Hölloch.

In May 1972, six years after the last attempt at the boulder choke, they returned to the clogged passage. Under the leadership of John Wilcox, a 35-year-old mechanical engineer from Columbus, Ohio, the Cave Research Foundation teams again attacked the exasperating wall of boulders. They hacked away for hours with digging tools, but got nowhere.

In July, however, team member Patricia Crowther found something new. A reed-thin physicist and the mother of two children, Crowther was already an experienced rock climber and caver when she joined the Flint Ridge group. Several hundred feet back from the rock pile, she crawled into a horizontal slot in an area that had not yet been thoroughly probed. After squirming about 100 feet, she came to a narrow, wedge-shaped crawlway, soon to be known as the Tight Spot. Pushing her back against its low ceiling so she would not slip into the V-shaped fissure beneath her, she hauled herself through and emerged at the top of a small waterfall that spilled over into a stand-up chamber. Her compass showed that the water was draining west toward Mammoth, and a lead was in sight. A few minutes later, she crawled back and rejoined the others. "It's very tight," she announced. "But we have cave!"

More than a month passed before Wilcox was able to send a small team back to look at Crowther's discovery. Veteran speleologist Roger Brucker, his 19-year-old son, Tom, and Flint Ridge newcomer Richard Zopf made their way to the Tight Spot. The elder Brucker was unable to get through, but Zopf and Tom Brucker made it. While Zopf caught his breath, Tom hurried through the stand-up room and slogged some 1,000 feet along the stream in a mud-walled passage. He felt certain that the stream emptied into the Mammoth Cave system, but he decided to wait for a full team before proceeding farther.

Four days later, Wilcox, Crowther, Zopf and Tom Brucker returned to

Mud-caked b[...]
that found th[...]
Ridge and Ma[...]
commemorati[...]
legendary cav[...]
foray Patricia[...]
had discovere[...]
passageway th[...]

By 1961, as shown in this relief view, cavers had explored elaborate networks of passages beneath the uplands of Flint and Mammoth Cave Ridges. But they knew of no route into the limestone bedrock of Houchins Valley.

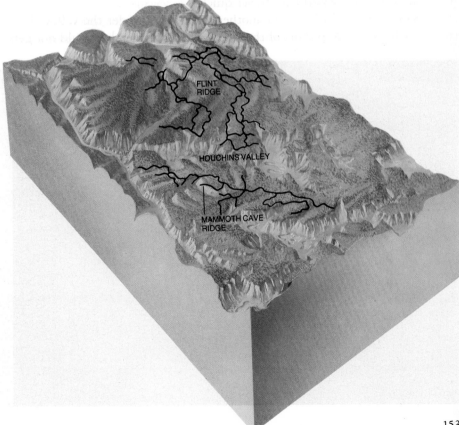

153

was flooded, as it is for much of the year. Almost certainly this explained how the route had been overlooked for so long. But later, when Stephen Bishop's map of 1842 was consulted, it showed this crucial passage had in fact been discovered by Mammoth's first and most famous guide.

As compelling as the adventure and romance of underground exploration may be, there are practical reasons for knowing the location and extent of caves. One of these was demonstrated by what happened to a cave just 10 miles east of the Flint Ridge system. Hidden River Cave, opened for tourism in 1916, boasted elegant galleries and offered scenic boat rides along the river's winding course through the cave. Hidden River also supplied drinking water to a small town situated above the cave; for decades the townspeople repaid this kindness by dumping sewage into nearby sinkholes, never suspecting that the sinkholes might feed back into the river. By the mid-1930s the water supply was hopelessly contaminated, and a decade later the tourist industry came to a halt as vile odors drove visitors back to the surface. The half-mile-long show cave and some 20 miles of lesser passages became a large sewer: The once-plentiful population of pearly blindfish in the river disappeared.

The Hidden River Cave disaster is by no means unique. Throughout the world, pollution from the surface poses a serious threat to caves—and thus to groundwater supplies at unexpected distances from the source of pollution. Karst regions are particularly susceptible to contamination because groundwater flows swiftly into the caves through joints in the rock instead of filtering slowly through purifying layers of sand and soil. The pollutants in the water typically sluice down into the cave's delicate ecosystem with all their raw toxicity intact. Cave creatures, which live precariously in the darkness, can be wiped out in a matter of hours.

Cave pollution has been a recognized problem ever since Édouard-Alfred Martel demonstrated in the 1890s that groundwater in karst areas carried disease-spreading bacteria and viruses. Martel's work to ban the dumping of wastes and animal carcasses into sinkholes undoubtedly saved thousands of lives in early-20th Century France, but underground water pollution still threatens residents of other European karst regions, especially in Austria, Yugoslavia and Italy, where large towns and cities tap subterranean streams for drinking water.

After World War II, the rapid rebuilding of European industry made cave pollutants even more deadly. Many industrial by-products are highly toxic and remain so for years; chemical wastes and heavy metals such as copper, mercury and chromium may poison the fragile underground environment for generations. In Yugoslavia, industrial pollution in karst regions annihilated more than 100 species of animal life in the famous Pivka underground river. More than a third of Austria is karstic, and in some places drinking water travels from a karst plateau to the household tap in less than half a day. Austria has designed special legislation to restrict development, industrialization and deforestation in karst regions.

A similar postwar industrial boom in the United States brought the same problems to American cave systems. And the pollution posed a frequent hazard for the steadily increasing number of amateur cavers. In 1966, the flame in a caver's carbide lamp ignited gasoline fumes in a cave about 100 miles northwest of Atlanta, Georgia, and the carbon monoxide fumes—formed by the incomplete combustion of the gasoline in the cave—killed three men. An investigation after the accident revealed that the gasoline had most likely seeped into the ground from a leak in a storage tank at a nearby service station. Three years later, a layer of gasoline seven feet thick

A map of the maze of the Flint Ridge and Mammoth Cave systems shows the route *(white)* of the expedition that proved they were connected. Although the charted passages appear to cross frequently, they actually run at many different depths and rarely intersect.

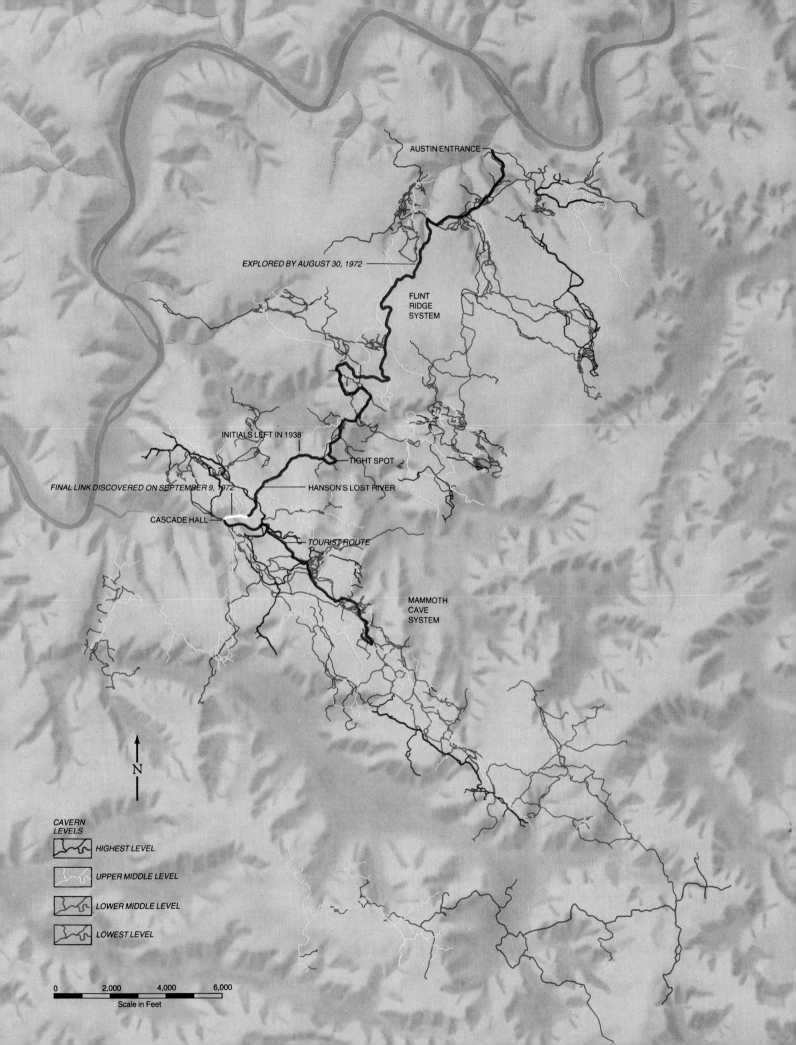

AUSTIN ENTRANCE

EXPLORED BY AUGUST 30, 1972

FLINT
RIDGE
SYSTEM

INITIALS LEFT IN 1938

TIGHT SPOT

FINAL LINK DISCOVERED ON SEPTEMBER 9, 1972

HANSON'S LOST RIVER

CASCADE HALL

TOURIST ROUTE

MAMMOTH
CAVE
SYSTEM

N

CAVERN
LEVELS

HIGHEST LEVEL

UPPER MIDDLE LEVEL

LOWER MIDDLE LEVEL

LOWEST LEVEL

0   2,000   4,000   6,000

Scale in Feet

was found floating on top of the groundwater in a well in a Pennsylvania karst area. The prime suspect was a local storage facility used by several oil companies. Almost one third of a square mile of groundwater was coated with gasoline; recovery crews collected more than 200,000 gallons.

As the world's karst areas become more populous, underground cavities can present another, even more direct threat to the surface dwellers. In a karst region in southern Italy, a house suddenly disappeared in 1978 when a cave roof collapsed beneath it, opening a new sinkhole. A parked car was sucked down with the house while the driver was away buying ice cream. In the United States, similar incidents occur fairly frequently in karst areas, where the land is honeycombed with subterranean cavities. In Alabama, 4,000 sinkholes have formed since 1900. Caused by drainage of underground water, these sinkholes have damaged highways, streets, railroads, buildings and pipelines. In central Florida, a sinkhole that developed in Winter Park in 1981 caused damage estimated at two million dollars.

Costly as these mishaps are, the effects could be far more disastrous if a sudden collapse destroyed a large dam. In any case, dams built on limestone are subject to other ills, including frequent leakage due to dissolution of the supporting rock. Repairs to dams in American karst regions have already cost tens of millions of dollars.

In an effort to detect subterranean cavities before major construction projects get started, scientists have developed a variety of geologic sensing devices. Among the most successful are underground radar, seismic detectors, acoustic resonance sensors and equipment that measures the change in the flow of an electric current around an underground cavity.

All these instruments rely on essentially the same principle—the measurement of changes in the flow of energy through the ground. Because the speed of energy traveling underground can be predicted accurately, changes in the rate can reveal the existence of a tunnel or a cave. For the seismic method, charges are detonated in boreholes to create measurable vibration. Radar waves transmitted through the ground between two sensors are equally predictable, and again, any interruption in the known pattern will indicate the presence of a cavity.

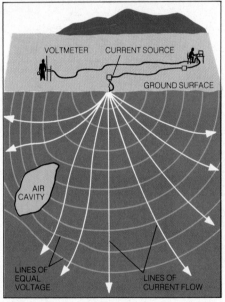

Technicians in a Texas field operate equipment that can locate underground voids by charting patterns of electrical resistivity. Current sent through the earth from a source electrode (*diagram*) loses voltage uniformly with distance unless an underground cavity disturbs the pattern; a computer can interpret the readings to plot the location, size and shape of the cavity.

158

To measure the extent of a cave that is already known, a high-intensity loudspeaker is sometimes placed inside the cave to blast sound into the cavity, causing the walls to resonate. This acoustic vibration travels through the rock walls to sensors on the surface. Analysis of the vibration then yields a statistical picture of all parts of the cave that connect directly to the sound source. For the detection of shallow cavities as small as several cubic feet, scientists measure the resistance of the soil to electrical energy. Electricity supplied by a portable generator is directed into the ground beneath a line of uniformly spaced electrodes. Because air is a poor electrical conductor, an empty cavity shows a significant resistance to the electrical current.

This new technology is showing promise for such applications as tracing karstic water supplies, and it might even be of use in finding new branches of major caves. But cavers by temperament tend to favor more traditional methods. Still working in teams and mapping carefully as they stoop, crawl and slither into new passages, explorers continue to push back the boundaries of the world's great caverns.

Kentucky cavers hardly paused for breath after the historic linkup of the Flint and Mammoth Cave Ridges. In 1979 the size of the Flint-Mammoth system was increased again when it was linked to the Joppa Ridge system southwest of Mammoth Cave. By 1982, the Flint-Mammoth-Joppa system had been extended to some 230 miles, and many tantalizing leads remained. "There is no doubt that it will reach 300 miles," said one Mammoth geologist, "and 500 is not impossible."

New records continue to be set in European karst regions as well. In 1981, a group exploring in the French Alps established a world depth record of 4,800 feet in a cave called Jean Bernard. After more than 30 years of exploration, Hölloch Cave in Switzerland is, at 87 miles, the third longest cave in the world, surpassed only by the Mammoth-Flint-Joppa system and the 89-mile Optimistichekaya Cave in the Ukraine.

Despite all the new detection technology, there is really no way to predict how many caves remain undiscovered. Most geologists agree that perhaps fewer than half of the caves in the United States have been located. With modern caving methods and equipment, it is difficult to imagine a cavern that could not be conquered by a crack team of speleologists. But in order to find these caves, explorers throughout the world will in the end rely on the most enduring caver's trait of all—passionate curiosity. Beyond that, they need a little luck and some well-earned experience. "I start with what I know and then poke around," said one veteran speleologist. "When I find a sinkhole I start digging." Ω

A "NEW CONTINENT" OF KARST

Despite the enthusiastic efforts of generations of cavers, only a small fraction of the earth's caves have so far been discovered, let alone explored. Yet while vast underground worlds remain hidden, speleologists have a good general idea of where they are. The caves lie beneath some of the earth's most distinctive and exotic terrain—areas of exposed limestone known as karst.

Named after the Karst area of Yugoslavia, where Serbian geographer Jovan Cvijič conducted the first comprehensive study of them in 1893, these landscapes are recognizable by the pronounced erosion of their highly soluble limestone. Rain water drains from other types of terrain into rivers and streams running through connected valleys. But rain falling on karst is absorbed almost immediately into the limestone's cracks, pits and sinkholes. Dissolution of the rock continues under the surface until networks of caves are carved out. The caves provide even better drainage and often cause rivers to disappear into the ground. Large depressions pockmark the karst landscape; the soil is thin or nonexistent; and, except in tropical regions, vegetation is sparse.

Karst regions occur throughout the world, notably in the Mediterranean basin, in parts of the Alps and the Pyrenees, in Kwangsi Province in southern China, and in Kentucky, Missouri and Tennessee in the United States. Local climatic conditions, especially rainfall, and details of geological history cause great variations in the appearance of karst, as the photographs on the following pages attest. The accompanying illustrations show where an experienced caver would go in pursuit of his ultimate achievement—the exploration of a previously unknown cave.

The frontier is so broad and the opportunities so immense that in 1966 the noted Swiss speleologist Alfred Bögli was moved to write: "Under the earth's crust there exists such an enormously great world, in absolute darkness, that we can with some justice speak of a new continent."

A karst formation called limestone pavement sprawls across the countryside in Yorkshire, England. The surface of the exposed limestone, scoured by glaciers during the last ice age, has been sculpted into deeply etched formations by water gradually eroding its joints.

This karst landscape, called a *polje,* from the
Serbo-Croatian word for "cultivated field," is in
the Taurus Mountains of southwest Turkey.
Melting snow floods the flat-floored valley each
spring; the rest of the year it is dry.

Most of the caves to be found in a *polje*
occur along its edges, because alluvial deposits
on the valley floor prevent absorption of
the seasonal flood and the water drains away at
the base of the limestone hills.

The limestone cliffs of a karst canyon loom
nearly 3,000 feet over a dry riverbed in Vikos
Gorge in northwest Greece. The river flows
aboveground in the wintertime, but during the
dry summer season it disappears into
underground cavities, emerging at the lower
end of the gorge as a great spring.

Cave systems are found in the walls of a karst
canyon at successive levels of the water table. As
the water table dropped, the older caves
were left high and relatively dry while the
cave-building process continued below.

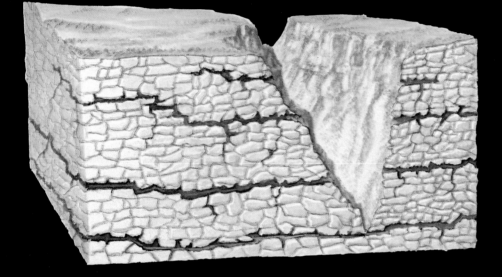

Depressions surrounded by cones of limestone
give cockpit karst, found only in the tropics,
its unique appearance and account for its name.
The jungle canopy of this Jamaican karst
region softens its contours, but conceals a relief
as jagged as that of other karst limestone.

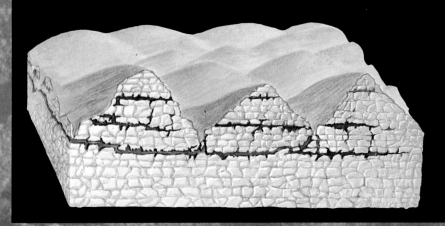

In cockpit karst, the dissolution of limestone is intensified by the carbonic acid generated by the decay of the abundant vegetation. Most cave entrances are found beneath the cockpits; older caves higher in the cones were formed along previous water-table levels.

An imposing array of 650-foot-high limestone
peaks overlooks fertile farmland in this
example of tower karst in southern China's
Kwangsi Province. Both the towers and
the flat valley floor below them are products of
thousands of years of erosion.

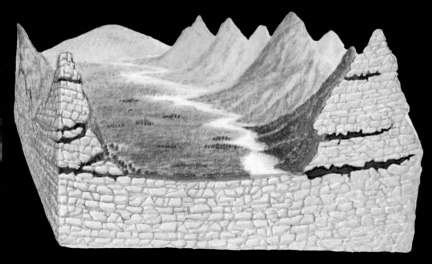

The ancient meandering river that is carving out
a cave at the base of a limestone tower (*right*)
formed the older, higher caves as it cut its way
downward. At left, accumulating rock debris
broadens a tower base to a more conical shape.

ACKNOWLEDGMENTS

For their help in the preparation of this book the editors wish to thank: **In Austria:** Vienna—Dr. Hubert Trimmel, University of Vienna. **In Belgium:** Chaudfontaine—Raymond Tercafs, Films Astrolabe. **In France:** Aven Armand—Jean Cabantous; Carmetin—Patrick Pallu; Moulis—Christian Juberthie, Laboratoire Souterrain; Nice—Michel Siffre; Paris—Claude Chabert; Jacques Choppy; Marcel Ichac; Haroun Tazieff; St.-Gaudens—Norbert Casteret; Valence—Pierre Agéron; Le Vésinet—Jacques Ertaut. **In Great Britain:** Leeds—Dr. Barry Webb, Institute of Geological Sciences; North Humberside—Andy Eavis; Nottingham—Dr. Tony Waltham, Trent Polytechnic. **In Italy:** Castellana—Nicola Mongelli, Grotte di Castellana. **In the People's Republic of China:** Beijing—Dr. Song Lin Hua, Institute of Geography, Academia Sinica; Karl-Heinz Bernhardt, Yu Ming-Bao, Shanthi Wikkranasinha, Li Xiaocong, Peking University; Zhao Fang, New China Picture Company. **In Switzerland:** Zurich—Alfred Bögli, Silva Verlag. **In the United States:** Alabama—(Huntsville) Jeanne Pritmore, The National Speleological Society; (Tuscaloosa) John G. Newton, Water Resources Division, United States Geological Survey; California—(Los Angeles) Dr. Clement Meighan, Department of Anthropology, University of California; Connecticut—(Storrs) Karen Kastning, Department of Geology and Geophysics, University of Connecticut; District of Columbia—Sarah and William Bishop, Cave Research Foundation; Duncan Morrow, Chief of Media Information, National Park Service; Indiana—(Marengo) Gordon L. Smith, President, National Caves Association; (Shelbeyville) Harold Meloy, Mammoth Cave Historian; Kentucky—(Louisville) Sharon Bidwell, Bernice Franklin, Librarians, *The Courier-Journal* and *The Louisville Times;* Tom Owen, Assistant Director, University Archives and Records Center, University of Louisville; (Mammoth Cave) Lewis D. Cutliff, Assistant Chief Park Interpreter, Mammoth Cave National Park; Massachusetts—(Boxboro) Ann Kress, Chairman, National Speleological Society Photo Archives; (Wilbraham) Emily Davis Mobley, Speleobooks; Missouri—(Protem) Tom Aley, Ozark Underground Laboratory; New Jersey—(Closter) Russell Gurney; New Mexico—(Carlsbad) Bobby L. Crisman, Ronal C. Kerbo, Carlsbad Caverns National Park; New York—(The Bronx) Brother G. Nicholas Sullivan, Manhattan College; Ohio—(Dayton) Roger W. Brucker; Pennsylvania—(Altoona) Jack H. Speece, Editor, *Journal of Spelean History;* (State College) Dr. and Mrs. William White; Tennessee—(Knoxville) Bill Deane; Texas—(Austin) William Mixon, Editor, *Speleo Digest;* (Dallas) Pete Lindsley, President, Richard Zopf, The Cave Research Foundation; (San Antonio) Thomas E. Owen, Southwest Research Institute; Virginia—(Arlington) Chip Clark; (Luray) Robert L. Bradford, Kermit B. Cavedo, Luray Caverns; H. T. N. Graves, President, Luray Caverns Corporation; (Reston) United States Geological Survey Library; (Richmond) John Wilson, Virginia Cave Commission; (Springfield) Paul Stevens, Executive Vice-President, National Speleological Society; Washington—(Seattle) Dr. William R. Halliday; Robert Stitt; Wisconsin—(Milwaukee) Dr. Merlin D. Tuttle, Curator of Mammals, Milwaukee Public Museum. **In Yugoslavia:** Ljubljana—Grozdana Kosak, Graficki Kabinet, Narodni Muzej; Postojna—France Habe; Srečko Šajn, Dino Borsellino, Postojna Jama.

The editors also wish to thank the following persons: Janny Hovinga, Wibo van de Linde, Amsterdam; Bob Gilmore, Auckland; Enid Farmer, Boston; Brigid Grauman, Brussels; Robert Kroon, Geneva; John Dunn, Melbourne; Dag Christensen, Oslo; Jimi Florcruz, Peking; Eva Stichova, Prague; Mary Johnson, Stockholm; Annelise Shulz, Vienna.

The index was prepared by Gisela S. Knight.

BIBLIOGRAPHY

Books

Ammen, S. Z., *The Caverns of Luray.* Allen, Lane & Scott's Printing House, 1886.

Bailey, Vernon, *Cave Life of Kentucky.* The University Press, no date.

Baker, Robin, ed., *The Mystery of Migration.* The Viking Press, 1981.

Balch, Edwin Swift, *Glacières or Freezing Caverns.* Johnson Reprint Corporation, 1970.

Balch, H. E., *Mendip: The Great Cave of Wookey Hole.* Bristol: John Wright & Sons, Ltd., 1947.

Baring-Gould, S., *The Deserts of Southern France.* Dodd, Mead & Co., 1894.

Barnett, John, *Carlsbad Caverns National Park, New Mexico.* Carlsbad Caverns Natural History Association, 1979.

Bauer, Ernst, *The Mysterious World of Caves,* London: Collins Publishers, 1971.

Bennett, Ross, ed., *The New America's Wonderlands: Our National Parks.* The National Geographic Society, 1980.

Bidegain, J., et al., *Marcel Loubens: Ses Souvenirs, Nos Témoignages.* Paris: Gallimard, 1958.

Binkerd, A. D., *The Mammoth Cave and Its Denizens: A Complete Descriptive Guide.* Robert Clarke & Co., 1869.

Bishop, Sherman C., *Handbook of Salamanders: The Salamanders of the United States, of Canada, and of Lower California.* Comstock Publishing Co., 1943.

Bögli, Alfred:
Féerie du Monde des Cavernes. Zurich: Editions Silva, 1976.
Karst Hydrology and Physical Speleology. Transl. by June C. Schmid. Springer-Verlag, 1980.

Boon, J. M., *Down to a Sunless Sea.* The Stalactite Press, 1977.

Borror, Donald J., *A Field Guide to the Insects of America North of Mexico.* Houghton Mifflin, 1970.

Brook, D. B., and A. C. Waltham, eds., *Caves of Mulu.* London: The Royal Geographical Society, 1978.

Brucker, Roger W., and Richard A. Watson, *The Longest Cave.* Alfred A. Knopf, 1980.

Bullit, Alexander Clark, *Rambles in the Mammoth Cave, during the Year 1844 by a Visitor.* Johnson Reprint Corporation, 1973.

Caiar, Ruth, and Jim White, Jr., *One Man's Dream.* Pageant Press, 1954.

Call, Ellsworth, *The Mammoth Cave.* Louisville & Nashville Railroad Co., 1856.

Casteret, Norbert:
The Darkness under the Earth. Henry Holt and Co., 1954.
The Descent of Pierre Saint-Martin. Transl. by John Warrington. The Philosophical Library Inc., 1956.
E.-A. Martel: Explorateur du Monde Souterrain. Paris: Gallimard, no date.
More Years under the Earth. Transl. by Rosemary Dinnage. London: Neville Spearman Ltd., 1961.
My Caves. Transl. by R. L. G. Irving. London: J. M. Dent & Sons, Ltd., 1947.
Ten Years under the Earth. Ed. and transl. by Barrows Mussey. Zephyrus Press, 1975.

The *Caves of the Earth: Their Natural History, Features, and Incidents.* American Sunday-School Union, 1953.

Charles, Jean-J., *Norbert Casteret.* Monaco: Éditions les Flots Bleus, 1958.

Chevalier, Pierre, *Subterranean Climbers: Twelve Years in the World's Deepest Chasm.* Transl. by E. M. Hatt. Zephyrus Press, 1976.

Clemens, Samuel, *Adventures of Huckleberry Finn.* Harper and Row, 1965.

Cullingford, C. H. D., *Exploring Caves.* Oxford University Press, 1951.

Darwin, Charles, *The Origin of Species by Means of Natural Selection.* Avenel Books, 1979.

De Joly, Robert, *Memoirs of a Speleologist; The Adventurous Life of a Famous French Cave Explorer.* Ed. by Pierre Boulanger. Transl. by Peter Kurz. Zephyrus Press, 1975.

Douglas, John Scott, *Caves of Mystery: The Story of Cave Exploration.* Dodd, Mead & Co., 1956.

Ellis, Bryan, *Surveying Caves.* Bridgewater, Somerset: The British Cave Research Association, 1976.

Eyre, Jim, *The Cave Explorers.* The Stalactite Press, 1981.

Fairbridge, Rhodes W., ed., *The Encyclopedia of Geomorphology,* Vol. 3. Dowden, Hutchinson & Ross, 1968.

Fairbridge, Rhodes W., and Joanne Bourgeois, eds., *The Encyclopedia of Sedimentology.* Dowden, Hutchinson & Ross, 1978.

Farr, Martyn, *The Darkness Beckons: The History and Development of Cave Diving.* London: Diadem Books Ltd., 1980.

Ford, T. D., and C. H. D. Cullingford, eds., *The Science of Speleology.* Academic Press, 1976.

Forwood, W. Stump, *An Historical and Descriptive Narrative of the Mammoth Cave of Kentucky.* J. B. Lippincott & Co., 1870.

Franke, Herbert W., *Wilderness under the Earth.* Transl. by Mervyn Savill. London: Lutterworth Press, 1958.

Gurnee, Russell H., *Discovery of Luray Caverns, Virginia.* R. H. Gurnee, Inc., 1978.

Gurnee, Russell and Jeanne, *Gurnee Guide to American Caves.* Zephyrus Press, 1980.

Habe, France, *The Postojna Caves.* Postojna: Postojnska jama, 1981.

Halliday, William R.:
   Adventure Is Underground. Harper & Brothers, 1959.
   American Caves and Caving. Harper & Row, 1974.
   Depths of the Earth: Caves and Cavers of the United States. Harper & Row, 1976.
Hanbury-Tenison, Robin, *Mulu: The Rain Forest.* London: Weidenfeld and Nicolson, 1980.
Herak, M., and V. T. Stringfield, eds., *Karst: Important Karst Regions of the Northern Hemisphere.* Elsevier Publishing Co., 1972.
Hill, Carol A., *Cave Minerals.* National Speleological Society, 1976.
Hogg, Garry, *Deep Down: Great Achievements in Cave Exploration.* Criterion Books, 1962.
Hovey, Horace Carter:
   Celebrated American Caverns. Johnson Reprint Corporation, 1970.
   Guide Book to The Mammoth Cave of Kentucky. Robert Clarke & Co., 1891.
Jakucs, László, *Morphogenetics of Karst Regions.* Transl. by B. Balkay. Bristol: Adam Hilger Ltd., 1977.
Jasinski, Marc, *La spéléologie.* Paris: Dargaud Éditeur, 1966.
Jennings, J. N., *Karst.* The M.I.T. Press, 1971.
Johnson, Peter, *The History of Mendip Caving.* Newton Abbot, Devon: David & Charles, 1967.
Judson, David, and Arthur Champion, *Caving and Potholing.* Granada Publishing Ltd., 1981.
Lawrence, Joe, Jr., and Roger W. Brucker, *The Caves Beyond.* Zephyrus Press, 1975.
Leroi-Gourhan, André, *Treasures of Prehistoric Art.* Harry N. Abrams, no date.
Lobeck, A. K., *Geomorphology: An Introduction to the Study of Landscapes.* McGraw-Hill, 1939.
Long, Abijah and Joe N., *The Big Cave.* Cushman Publications, 1956.
Lovelock, James, *Caving.* London: B. T. Batsford Ltd., 1969.
McClurg, David R., *Exploring Caves: A Guide to the Underground Wilderness.* Stackpole Books, 1980.
Martel, E.-A.:
   Les Abîmes: Les Eaux Souterraines, Les Cavernes, Les Sources. Paris: Librairie Charles Delagrave, 1894.
   L'Aven Armand: Description Géologie Historique. Millau: Editions Artières, 1962.
Meloy, Harold, *Mummies of Mammoth Cave.* Micron Publishing Co., 1977.
Minvielle, Pierre, *Grottes et Canyons: Les 100 Plus Belles Courses et Randonnées.* Paris: Éditions Denoël, 1977.
Mohr, Charles E., *The World of the Bat.* J. B. Lippincott, 1976.
Mohr, Charles E., and Thomas L. Poulson, *The Life of the Cave.* McGraw-Hill, 1966.
Mohr, Charles E., and Howard N. Sloane, eds., *Celebrated American Caves.* Rutgers University Press, 1955.
Moore, George W., and Brother G. Nicholas, *Speleology: The Study of Caves.* D. C. Heath and Co., 1964.
Murray, Robert K., and Roger W. Brucker, *Trapped!* G. P. Putnam's Sons, 1979.
Nicod, Jean, *Pays et Paysages du Calcaire.* Paris: Presses Universitaires de France, 1972.
Notice sur les Travaux Scientifiques de M. Édouard-Alfred Martel. Paris: Libraires de l'Académie de Médecine, 1911.
Owen, Luella Agnes, *Cave Regions of the Ozarks and Black Hills.* Johnson Reprint Corporation, 1970.
Palmer, Arthur N., *A Geological Guide to Mammoth Cave National Park.* Zephyrus Press, 1981.
Partington, J. R., *A Short History of Chemistry.* Harper & Brothers, 1960.

Šajn, Srečko, ed., *Postojnska Jama.* Transl. by Zdenka Šlenc. Postojna: Postojnska jama THO, 1978.
Scheffel, Richard L., and Susan J. Wernert, eds., *Natural Wonders of the World.* The Reader's Digest Association, 1980.
Shaw, Trevor R., *History of Cave Science.* Crymych, Wales: Anne Oldham, 1979.
Sieveking, Ann and Gale, *The Caves of France and Northern Spain: A Guide.* London: Vista Books, 1962.
Siffre, Michel:
   Dans les Abîmes de la Terre. Paris: Flammarion, 1975.
   Grottes, Gouffres & Abîmes. Paris: Hachette Réalités, 1981.
Siffre, Michel, and Georges Dupont, *Les Animaux de Gouffres et des Cavernes.* Paris: Hachette, 1979.
Sloane, Bruce, ed., *Cavers, Caves and Caving.* Rutgers University Press, 1977.
Sparks, B. W., *Geomorphology.* London: Longmans, Green and Co., Ltd., 1960.
Stenuit, Robert, and Marc Jasinski, *Caves and the Marvellous World beneath Us.* A. S. Barnes and Co., 1966.
Sweeting, Marjorie M., *Karst Landforms.* Columbia University Press, 1973.
Taylor, Bayard, *At Home and Abroad.* G. P. Putnam, 1862.
Thompson, Ralph Seymour, *The Sucker's Visit to the Mammoth Cave.* Johnson Reprint Corporation, 1970.
Verne, Jules, *Journey to the Center of the Earth.* Dodd, Mead & Co., 1979.
Waltham, A. C., *The World of Caves.* London: Orbis Publishing Ltd., 1976.
Waltham, Tony, *Caves.* Crown Publishers, 1977.
Weaver, Dwight H., *Onondaga: The Mammoth Cave of Missouri.* Discovery Enterprises, 1973.
Yalden, D. W., and P. A. Morris, *The Lives of Bats.* Quadrangle/The New York Times Book Co., 1975.

Periodicals
The British Caver. Vols. 67-84, Christmas 1977-Spring 1982.
Bulletin of the National Speleological Society. June 1940-December 1981.
Burman, Ben Lucien, "Kentucky's Crazy Cave." *Collier's,* June 6, 1953.
Casteret, Norbert, "Discovering the Oldest Statues in the World." *National Geographic,* August 1924.
"Caves and Caving." *The Bulletin of the British Cave Research Association,* Nos. 1-13, August 1978-August 1981.
Caving International Magazine. Nos. 1-12, October 1978-July 1981.
Crowther, Patricia:
   "Discovering the World's Biggest Cave." *The Saturday Review,* April 1973.
   "Into Mammoth Cave—The Hard Way." *National Parks & Conservation Magazine,* January 1973.
Detjen, Jim, and Jim Adams, "Hidden River Cave Was Unable to Hide from Man's Pollution." *The Courier-Journal* (Louisville), December 2, 1979.
Durand, J. P., and A. Vandel, "Proteus: An Evolutionary Relic." *Science Journal,* February 1968.
GEO² (Publication of Cave Geology and Geography Section at the National Speleological Society), 1974-1981.
Gilbert, Bil, "Batty about Caves." *Sports Illustrated,* March 15, 1982.

Griffin, Donald R., "Mystery Mammals of the Twilight." *National Geographic,* July 1946.
Hapgood, Fred, "The Ghostly Wings of Night." *GEO,* July 1981.
Herald, Earl S., "Texas Blind Salamander in the Aquarium." *Aquarium Journal,* August 1952.
The Journal of Spelean History, 1968-1981.
Kurtén, Björn, "The Cave Bear." *Scientific American,* March 1972.
Lee, Willis T.:
   "New Discoveries in Carlsbad Cavern." *National Geographic,* September 1925.
   "A Visit to Carlsbad Cavern." *National Geographic,* January 1924.
Lesy, Michael, "Dark Carnival: The Death of Floyd Collins." *American Heritage,* October 1976.
Marshack, Alexander, "Exploring the Mind of Ice Age Man." *National Geographic,* January 1975.
Martel, E.-A.:
   "British Caves and Speleology." *The Geographic Journal,* July-December 1897.
   "The Descent of Gaping Ghyll (Yorkshire): A Story of Mountaineering Reversed." *The Alpine Journal,* May 1896.
   "Into the Earth's Depths: Twenty Years of Cave-Exploring." *Sunday Magazine,* January 28, 1906.
   "The Land of the Causses: The Canon of the Tarn, Montpellier-le-Vieux." *Appalachia: The Journal of the Appalachian Mountain Club,* 1893-1895.
   "Speleology: A Modern Sporting Science," *The Yorkshire Ramblers' Club Journal,* 1903-1908.
   "Speleology, or Cave Exploration." *Appleton's Popular Science Monthly,* December 1898.
Mohr, Charles E.:
   "Exploring America Underground." *National Geographic,* June 1964.
   "I Explore Caves." *Natural History,* April 1939.
   "Ozark Cave Life." *National Speleological Society Bulletin,* November 1950.
Nicholas (Sullivan), Brother G., "Entrance, Twilight and Dark." *Natural History,* April 1971.
NSS News, January 1941-April 1982.
Oster, Gerald, "The Modern Look of Ice Age Art." *Natural History,* October 1978.
Poulson, Thomas L., "Cave Adaptation in Amblyopsid Fishes." *The American Midland Naturalist,* 1963.
Poulson, Thomas L., and William B. White, "The Cave Environment." *Science,* September 5, 1969.
The Quarterly Journal of the Geological Society of London, 1965.
Ross, Edward S., "Birds That 'See' in the Dark with Their Ears." *National Geographic,* February 1965.
Sainte-Croix, L. de, "E. A. Martel: Explorateur des Abîmes et des Eaux Souterraines." *Revue de l'Alliance Française,* April 1933.
Schreiber, Richard, "The Disaster at Howard's Waterfall Cave." *Georgia Underground,* March-April 1966.
Shields, Mitchell J., "The Lure of the Abyss." *GEO,* June 1979.
Speleo Digest, 1974 and 1978.
Woods, Loren P., "Blind Fishes Found in Cave Pools and Streams." *National Speleological Society Bulletin,* December 1956.
Yager, Jill, "Remipedia, A New Class of Crustacea from a Marine Cave in the Bahamas." *Journal of Crustacean Biology,* 1981.

Other Publications
Barnett, John, "Carlsbad Caverns National Park, New Mexico." Carlsbad Caverns Natural History Association pamphlet, 1979.
Beck, Barry F., ed., *Proceedings of the Eighth International Congress of Speleology,* Vols. 1 and 2. Meet-

ing of the International Union of Speleology, Bowling Green, Kentucky, July 18-24, 1981.

Davies, W. E., and I. M. Morgan, "Geology of Caves." U.S. Department of the Interior Geological Survey pamphlet.

"Earth Resistivity Exploration." Southwest Research Institute pamphlet, San Antonio, Texas.

Eavis, A. J., ed., *Caves of Mulu '80: The Limestone Caves of the Gunong Mulu National Park, Sarawak.* London: The Royal Geographical Society, 1981.

Monroe, Watson H., "A Glossary of Karst Terminology." Geological Survey Water-Supply Paper

1899-K. United States Printing Office, 1970.

Parzefall, Jakob, Jaques Durand and Bernard Richard, "Aggressive Behavior of the European Cave Salamander *Proteus anguinus.*" Paper presented at the Eighth International Congress of Speleology, Bowling Green, Kentucky, 1981.

Peters, Wendell R., and Richard G. Burdick, "Use of an Automatic Earth Resistivity System for Detection of Abandoned Mine Workings." Society of Mining Engineers of the AIME Preprint No. 81-89. Paper presented at the AIME Annual Meeting, Chicago, February 22-26, 1981.

Webb, Barry, and Antony C. Waltham, unpublished manuscript on Mulu National Park.

White, Jim, "Carlsbad Caverns National Park, New Mexico: Its Early Explorations as Told by Jim White." Privately published pamphlet.

White, William, ed., "Geology and Biology of Pennsylvania Caves." General Geology Report 66, Pennsylvania Geological Survey, Fourth Series, Harrisburg, 1976.

Wilford, G. E., *The Geology of Sarawak and Sabah Caves.* Geological Survey, Borneo Region, Malaysia, Bulletin 6, 1964.

PICTURE CREDITS

The sources for the illustrations that appear in this book are listed below. Credits from left to right are separated by semicolons, from top to bottom by dashes.

Cover: © Michael K. Nichols from Woodfin Camp & Associates. 6, 7: Philippe Crochet, Blois, France. 8: Chip Clark. 9: Francis Le Guen, Paris. 10, 11: Ron Simmons—© 1978 Michael K. Nichols from Woodfin Camp & Associates. 12, 13: © Jay Lurie from Bruce Coleman, Inc.; © C. B. Frith from Bruce Coleman, Inc. 14, 15: Dr. Arthur N. Palmer. 16: Courtesy Austrian National Library, Vienna. 18: Map by Bill Hezlep and Jaime Quintero. 20, 21: Art by Bob Wood. 22: Colin Boothroyd, courtesy Andy Eavis, North Humberside, England. 23: Art by I'Ann Blanchette. 24, 25: Colin Boothroyd, courtesy Andy Eavis, North Humberside, England. 26: Photo Bulloz, courtesy Musée du Petit Palais, Paris. 27: Michael Holford, courtesy Victoria and Albert Museum, London. 29: Jean Vertut, Issy-les-Moulineaux, France. 30, 31: Jean Vertut, Issy-les-Moulineaux, France (2)—Des and Jen Bartlett from Bruce Coleman, Inc.; Ernst H. Kastning; Jean Vertut, Issy-les-Moulineaux, France. 32, 33: Courtesy Austrian National Library, Vienna. 34: By permission of the Houghton Library, Harvard University. 36-40: Library of Congress. 42-49: Joseph Natanson, courtesy Narodnega Muzeja, Ljubljana, Yugoslavia. 50: Scott Lamb. 52: Boyer-Viollet, Paris. 53: Library of Congress. 55: David E. Bunnell. 56, 57: © 1981 Tom Bean—Grant Heilman Photography; Ross Ellis, Sydney. 60:

Mary Evans Picture Library, London. 61: Art by I'Ann Blanchette. 62, 63: Library of Congress. 64, 65: Map by Bill Hezlep. 66: David E. Bunnell. 67: Art by I'Ann Blanchette. 68-70: Art by Jaime Quintero. 72: Ernst H. Kastning. 73: Y. and F. M. Callot, Vic-en-Bigorre, France. 75: © 1981 Jim Tuten from Black Star. 76, 77: Yoram Lehmann from Robert Harding Picture Library, London. 78: © 1977 Kenrick L. Day. 79: National Park Service. 80, 81: Elaine Garifine (2); Jacques Choppy, Paris. 82: Dr. Arthur N. Palmer—© 1977 Kenrick L. Day. 83: Philippe Crochet, Blois, France. 84: Luray Caverns, Virginia—Chip Clark—Shostal Associates. 85: Ronal C. Kerbo. 86: Collection LA VIE DU RAIL, © P. L. M. 1982, Paris. 88, 89: National Park Service. 90, 91: Art by Lloyd K. Townsend. 93: © 1973 Russ Kinne from Photo Researchers Inc.—The Cavern Supply Company, Inc. 94, 95: The Cavern Supply Company, Inc., except top right, Ed Cooper. 96: Courtesy Gordon L. Smith. 97: National Park Service/Mammoth Cave National Park. 98: Courtesy Gordon L. Smith. 100: UPI. 101: Art by I'Ann Blanchette. 102: Wide World, courtesy *The New York Times.* 103: Reprinted with permission from *The Courier-Journal* and *The Louisville Times.* 104, 105: Courtesy OEFVW, Vienna. 106: Art by I'Ann Blanchette. 107: © *Illustrated London News,* London. 108: Daniel Roucheux, Montrouge, France. 109: Y. and F. M. Callot, Vic-en-Bigorre, France. 110-117: Art by Jack Unruh. 118: Chip Clark. 120: Art by Frederic F. Bigio from B-C Graphics. 123: Charles E. Mohr. 124,

125: Charles E. Mohr, except far right, Scott Lamb. 126: J. A. L. Cooke from Oxford Scientific Films, Oxford, England. 128: ANIMALS ANIMALS/© Raymond A. Mendez. 130: Scott Lamb. 131: ANIMALS ANIMALS/© Robert W. Mitchell. 132: Tercafs—Jacana, Paris; Library of Congress. 133: Charles E. Mohr—© Marvin E. Newman. 134: © Brian Brake from Photo Researchers Inc. 135: Dennis W. Williams. 136, 137: Chip Clark. 138, 139: Charles E. Mohr; Arlan R. Wiker. 140: Charles E. Mohr. 141: ANIMALS ANIMALS/© Robert W. Mitchell. 142: Photo Le Parisien Libéré, Saint-Ouen, France. 144: Jacques Ertaud, courtesy Éditions Arthaud, Paris. 147: Art by I'Ann Blanchette. 148: Photo Le Parisien Libéré, Saint-Ouen, France. 149: *Paris-Match*/Saulnier, Paris. 151: © Martyn Farr from All Sport/Photographic Ltd., Morden, Surrey, England—Martyn Farr, Powys, Wales. 152: Reprinted with permission from *The Courier-Journal* and *The Louisville Times.* 153: Art by Jaime Quintero. 154, 155: David DesMarais/CRF. 157: Map by Bill Hezlep and John Britt. 158: Southwest Research Institute, San Antonio, Texas—art by Frederic F. Bigio from B-C Graphics. 160, 161: Ernst H. Kastning. 162, 163: John Middleton, Sheffield, England, inset, art by Jaime Quintero. 164: Jean-Pierre Charre, Grenoble, France. 165: Art by Jaime Quintero. 166, 167: Anthony Waltham, Nottingham, England, inset, art by Jaime Quintero. 168, 169: © 1979 George Holton from Photo Researchers Inc., inset, art by Jaime Quintero.

INDEX

Printed and bound in Italy by
A. Mondadori, Verona.